KT-239-016

# AFFORESTATION

# AFFORESTATION

## Policies, planning and progress

Edited by Alexander Mather

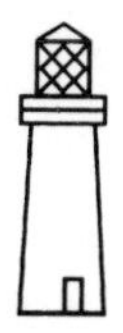

**Belhaven Press**
**London and Florida**

**Belhaven Press**
(a division of Pinter Publishers)
25 Floral Street, Covent Garden, London, WC2E 9DS, United Kingdom

First published in 1993



**British Library Cataloguing in Publication Data**

A CIP catalogue record for this book is available from the British Library

ISBN 1 85293 202 3

Distributed in North America by
CRC Press, 2000 Corporate Blvd., N.W.,
Boca Raton, Florida, 33431

**Library of Congress Cataloging-in-Publication Data**
A CIP catalog record for this book is available from the Library of Congress

Typeset by Koinonia Ltd, Manchester
Printed and bound in Great Britain by Biddles Ltd, Guildford and King's Lynn

# Contents

# List of figures

# List of tables

# List of contributors

**Professor Yoshihisa Fujita**, Department of Geography, Aichi University, Machihata-machi, Toyohashi City 441, Japan

**Professor Desmond Gillmor**, Department of Geography, Trinity College, Dublin 2, Ireland

**Dr Helen Groome**, Colon de Larreategui 31–3, Bilbao 48009, Spain

**Lektor Kr. M. Jensen**, Tuborgvej 153, DK 2900 Hellerup, Denmark

**Professor Bela Keresztesi**, Forest Research Institute, 1023 Budapest Frankel Leo U. 42–44, Budapest, Hungary

**Mr Antonio Lara**, Institute of Tree Ring Research, University of Arizona 85721, USA

**Dr A. S. Mather**, Department of Geography, University of Aberdeen, Aberdeen AB9 2UF, UK

**Dr Geoff McDonald**, Division of Australian Environmental Studies, Griffith University, Nathan, Brisbane, Queensland, Australia

**Professors Michael Roche** and **Richard Le Heron**, Department of Geography, Massey University, Palmerston North, New Zealand

**Professor Vaclav Smil**, Department of Geography, University of Manitoba, Winnipeg, Manitoba R3T 2N2, Canada

**Mr Philip Stewart**, Oxford Forestry Institute, South Parks Road, Oxford OX1 3RB, UK

**Professor T. Veblen**, Department of Geography, University of Colorado at Boulder, 110 Guggenheim, Campus Box 260, Boulder, Colorado 80309-0260, USA

**Dr Michael Williams**, School of Geography, University of Oxford, Mansfield Road, Oxford OX1 3TB, UK

# 1 Introduction

*Alexander Mather*

> 'Saving the forests ... is not just a matter of replanting trees, but it would be a good start' (Braden 1991, p. 130).

This statement was made about the forests of the former Soviet Union, but it is equally pertinent to the global forests. Saving the forests has become a major concern in the late twentieth century, not only for their resource value but also for the environmental benefits they provide. It is a truism that the forests cannot be saved simply by afforestation. Indeed, in some circumstances afforestation may even be a threat to forests, if, for example, single-species industrial plantations replace native woodlands. Nevertheless, it may also be a means of restoring forest cover in areas that have suffered deforestation, and thus it can help to maintain the global forest area.

Much has been written about deforestation and its causes and possible remedies. Far less attention has been focused on afforestation. The aim of this volume is to review achievements in afforestation in a variety of national settings, and in particular to examine afforestation policies, planning and progress. Where possible, the origins of afforestation policies are identified, and the problems and difficulties that have arisen as they have been implemented are discussed. In a number of countries, progress in afforestation during the present century has been substantial, to the extent that the forest area has more than doubled. Almost invariably, however, problems have been encountered, either in implementing policies and achieving planting targets, or in the associated environmental or social effects. As more and more countries seek to develop and implement afforestation programmes, the shared experiences of different lands may offer useful lessons, even if it is the case that they can rarely be transferred directly from one setting to another.

## Afforestation and reforestation

Afforestation is difficult to define in precise and rigorous terms. Whilst a literal meaning may be simply to 'convert into forest', such a process is unlikely to occur on land which at some time in the past has not borne forest. Afforestation and reforestation are therefore not easily distinguished except in rather arbitrary terms of the number of years during which the land has been bare. Comparative statistical data on forest areas in individual countries are scarce enough: comprehensive data on rates and extents of precisely defined afforestation at the international or global scales are simply not available.

An apparently simple distinction can be made between afforestation and reforestation. The former term usually applies to land that has lacked forest for several decades or centuries, whereas reforestation relates to the more immediate restoration of forest cover. The apparent simplicity of this distinction, however, is misleading: many more shades of meaning can exist. Following an FAO world symposium on 'man-made' forests in the 1960s (FAO 1967), Evans (1992) distinguishes five categories:

(i) *afforestation* of bare land, which has carried no forest for at least 50 years;
(ii) *reforestation* of land that has carried forest within 50 years but where the previous crop is replaced by an essentially different one;
(iii) *reforestation*, establishing essentially the same crop as before;
(iv) forests established by natural regeneration assisted by silvicultural intervention;
(v) forests regenerated naturally without human assistance.

Conceptually, these categories are distinct and precisely defined, but in practice few if any sets of national statistics relate clearly to such terms. The '50-year' distinction between afforestation and reforestation is especially problematic, as is the distinction between the two types of reforestation.

These categories are defined on the basis of two main criteria – the duration of absence of forest, and the means of re-establishing tree cover. In theory a contrast can be drawn between the active creation of new forests through planting, and the passive regeneration of forest on land abandoned by agriculture. Whether spontaneous regeneration occurs, however, may depend on factors such as the survival of local seed sources and the extent to which the abandoned land is capable of sustaining the growth of seedlings. Such a contrast may be significant in terms of the cost and difficulty of re-establishing forest cover, but essentially the same end result is achieved in both cases. Again, the contrast is rarely drawn in national or international statistics.

Afforestation is generally viewed in this volume as the establishment of forest, by planting or regeneration, on land recently lacking forest cover. For some countries such a definition gives rise to few problems, but in other cases it is almost impossible, on the basis of existing statistics and policies, to distinguish meaningfully between afforestation and reforestation. ('Reafforestation' is sometimes used to cover both processes.) The variable degree of clarity between the terms in different countries is apparent in the ensuing chapters.

A further problem that becomes increasingly apparent is that of the availability and reliability of statistics on forest areas and their trends. Many of these problems are indicated in statistical compilations published annually by bodies such as FAO, UNEP and WRI, and general problems of forest statistics are reviewed elsewhere (e.g. Williams 1989a; Mather 1990). The inadequate and unsatisfactory nature of statistical coverage of forest areas in general and of rates of afforestation and reforestation in particular cannot be too strongly emphasised. The most comprehensive and convenient statistical source on forest area is the annual FAO *Production Yearbook*. In it, '"forests and woodland" refers to land under natural or planted stands of trees, whether productive or not, and includes land from which forests have been cleared but which will be reforested in the foreseeable future'. The latter part of this definition obviously poses major problems, over and above those arising from the definition of 'forest' itself. A further problem is that afforestation achievements reported in the literature can be misleading. Even if the afforested *area* is correctly stated, survival rates may be low, to the extent that two or more hectares are planted for each hectare of forest successfully established.

## Afforestation: historical background

The technical ability to plant trees has existed for at least two thousand years. Large-scale afforestation was practised in parts of the ancient Mediterranean world, with trees being transplanted from nurseries where they were raised from seed or cuttings. Both timber-producing and crop-producing trees were planted: trees other than olives and vines were planted at least as early as 255 BC (Glacken 1967). One Roman writer in the first century AD remarked that the mountain slopes around his villa were 'covered with plantations of timber' (Hughes with Thirgood 1982). In the Orient, commercial tree planting in China is recorded prior to the sixth century AD (Richardson 1990). In Europe, the planting of conifers to fix sand dunes is recorded in Portugal from as early as the fourteenth century, and in subsequent centuries extensive areas were planted on similar sites on the Atlantic coast of France and along the shores of the Baltic (Glacken 1967). By the seventeenth and eighteenth centuries, afforestation was being carried out in Japan both for timber production and for environmental management (Totman 1986). With increasing shortages of oak for shipbuilding, countries such as Britain and the Netherlands began to look towards teak plantations in India and the East Indies respectively, as well as attempting to safeguard home supplies of timber.

Despite its long history, the scale of afforestation generally remained small until recently. Sustained, large-scale afforestation was not widespread until the twentieth century in Europe, and it has not spread extensively beyond Europe until the last quarter of this century.

## Forest characteristics

The products of afforestation usually differ from natural forests in a variety of respects. A greater degree of uniformity is likely to occur. The range of tree species is usually much narrower than in natural forests, and extensive stands of uniform age-classes are typical. Many plantations are composed mainly of one or two species, and in modern plantations the species are often exotic rather than native. Afforestation may restore a form of forest cover, but is most unlikely to re-create the conditions of the natural forest. In the literal sense of restoring the orginal forest conditions, therefore, reforestation is something of a misnomer, although some types of afforestation tend more closely to natural conditions than others.

Forests are sometimes established for reasons of environmental protection rather than primarily for wood production, and therefore their productivity is not always of great significance. Nevertheless, it is a often a characteristic of major importance. Planted forests are usually much more productive, in terms of increment of wood per unit area and time, than natural forests. And plantations may also offer the advantage of greater uniformity of type of wood and greater proportions of potentially useful stemwood than is to be found in natural forests.

Productivities are often especially high in tropical and sub-tropical plantations. Reports of some very high productivities achieved in tropical plantations need to be viewed with caution. In some cases they may be from experimental plots rather than from large forests, and they may also include bark. Nevertheless, they are typically many times higher than those achieved in more traditional forest lands, and may also offer the possibility of much shorter rotations. For example mean annual increments may range from 1 to 3 $m^3$/ha per year over rotations of the order of 100 years in areas such as Canada and Scandinavia to more than 20 $m^3$/ha per year over 20–25 year rotations in pine plantations in countries such as Chile and New Zealand. Even higher productivities and shorter rotations may be possible in more tropical latitudes (e.g. Evans 1982; Sedjo 1984).

Because of their high productivities, planted forests usually make contributions to industrial wood production that are quite disproportionate to their area. In Latin America, for example, plantations in the mid-1980s comprised 0.6 per cent of the forest area, yet produced 30 per cent of the industrial wood supply (McDonald and Krugman 1986). In Australia, the corresponding percentages are 0.7 and 54 (Turner and Gessel 1990). The trend is clear: world wide, increasing proportions of industrial wood are coming from planted forests. As the area of plantations increases and as native forests are increasingly protected for their environmental values, the trend will become even more apparent.

## Forests and plantations

The world forest area extends to around 4 billion ha, made up of approximately 2.8 billion and 1.2 billion ha respectively of closed and open forest. The precise extent of 'man-made' forests or plantations is uncertain, but is likely to be between 125 and 150 million hectares (Mha), or 3–4 per cent of the global forest area. The plantation area was estimated to be around 81 Mha in 1965 (Logan 1967), and in the early 1970s may have been approximately 95 Mha (Persson 1974). More recently Evans (1986) estimates the area to be between 120 and 140 Mha. The same author suggests that the area of plantations in the tropics (defined as the zone between 27°N and 27°S) was approximately 43 Mha in 1990 (Evans 1992): in other words, at least two-thirds of the total area is in temperate latitudes. By far the greater part of the plantation area is in the Northern Hemisphere. The total area in the Southern Hemisphere has recently been estimated at 8.3 Mha (Turner and Gessel 1990).

Most of the afforested area is geared to the production of industrial wood rather than fuelwood. Industrial plantations were indicated by Postel and Heise (1988) as extending to nearly 92 Mha in 1985. Of this area, most lay in the temperate zone.

Compared with the estimated extent of plantations, recent rates of planting are high and are accelerating. Afforestation – or the creation of plantations – has tended to increase in many countries, and to spread to far more countries than were involved during the first half of this century. In the United States, for example, annual planting increased sixfold from 200,000 ha in 1950 to 1,200,000 ha in 1987 (Sedjo 1991). In Thailand, annual rates of 'reforestation' increased from 2000 ha in the early 1960s to 39,500 ha in the early 1980s (Kunstadter 1990). This led to a growth in the plantation area from 21,000 ha in 1965 to 560,000 ha in 1990 (Evans 1992). In Brazil, the area of plantations increased from 0.5 to 7.15 Mha between 1965 and 1990, with a target of 12 Mha by AD 2000 (Evans 1992). The same author estimates expansion of plantations in India from 0.95 Mha in 1965 to 14 Mha in 1990. A similar story of increasing afforestation, albeit on a smaller scale, applies to Africa (Grainger 1986).

In the 15 years up to 1980, 14 Mha of plantations are estimated to have been established in the tropics, whereas between 1980 and 1990 the annual rate was one-and-a-half times as high at nearly 2 Mha (Evans 1992).

UNEP (1991) indicates that the annual rate of 'reforestation' in the late 1980s was around 14.7 Mha. Such a figure is, of course, subject to the qualifications previously expressed about the reliability of statistics, and in particular about survival rates. On paper, however, it is not far short of the annual 'deforestation' rate of 15.5 Mha reported by UNEP (1991), although deforestation rates may be understated and afforestation rates overstated in official statistics. It should also be emphasised that 'reforestation' and planting are not necessarily synonymous. Savill and Evans (1986) have assembled statistics of 'annual forest renewal' and 'annual planting' for a variety of countries. While in some cases the rates are identical, in others the former is much higher than the latter.

Afforestation and deforestation do not coincide spatially at the global scale (or indeed at more local levels), and parity of rates would not necessarily mean stability of forest areas. Nevertheless rates of 'reforestation' have probably increased considerably in recent years: the 1988–89 and 1990–91 volumes of *World Resources*, for example, suggest that reforestation in South America has increased from 416,000 to 760,000 ha and in Africa from 196,000 to 355,000 ha.

Even allowing for statistical deficiencies, it seems clear that the global gap between rates of deforestation and reforestation has narrowed. In Africa, for example, the ratio of planting to deforestation is given by Lanly (1982) as 1:29, compared with the recent ratio of deforestation to reforestation as quoted by UNEP (1991) of 1:14. Lanly suggests that the tropical Asian ratio was 1:4.5, whereas UNEP statistics suggest that 'reforestation' now exceeds deforestation by a margin of 25 per cent in Asia. Care is required in reaching conclusions on the basis of such data, especially since Asian trends are strongly influenced by the case of China (Chapter 8). Part of these increases in reported 'reforestation' rates may result from improved reporting and statistical coverage, but there nevertheless appears to be a real underlying trend. It remains to be seen whether 'reforestation' rates will match those of 'deforestation' at the global scale in the near future, and whether the global forest area will stabilise or even increase.

## Afforestation and the forest transition

A major contrast exists between forest trends in the developed and developing worlds. This contrast is epitomised by the fact that the percentage of the European

**Table 1.1** Changes in reported area of forest and woodland

| | Area of forest and woodland as percentage of land area | | |
|---|---|---|---|
| | 1966–68 | 1986–88 | change |
| Britain | 8 | 10 | +2 |
| Ireland | 3 | 5 | +2 |
| Denmark | 11 | 12 | +1 |
| Hungary | 16 | 18 | +2 |
| Spain | 27 | 31 | +4 |
| Algeria | 1 | 2 | +1 |
| China | 16 | 14 | –2 |
| Chile | 12 | 12 | 0 |
| New Zealand | 27 | 27 | 0 |
| Australia | 18 | 14 | –4 |
| Japan | 67 | 67 | 0 |
| USA | 32 | 29 | –3 |
| World | 33 | 31 | –2 |
| Africa | 25 | 23 | –2 |
| Europe | 31 | 33 | +2 |
| OECD Europe | 32.8 | 33.1 | +0.3 |
| OECD* | 33.2 | 33.2 | 0 |

*OECD figures are for 1980–89

*Sources*: FAO *Production Yearbooks* (various years); UNEP (1991); OECD (1991).

land area under forest increased by two points during the twenty years up to the late 1980s, while that in Africa declined by 2 per cent (Table 1.1).

Many parts of the developing world are suffering rapid deforestation, and the fate of the tropical forest in particular is a major global environmental issue. In much of the developed world, however, the forest area is clearly expanding. Levels of population and demand for food are growing only slowly if at all, and advances in agricultural technology have allowed yields to increase rapidly in recent decades. Land has therefore been released from agriculture, especially in areas where it is not of prime quality. Marginal areas, perhaps colonised by agriculture in the face of rapid population growth during the nineteenth century, may be abandoned as their populations drift to the cities or turn their attention to activities such as tourism. Part of the area thus released is now under forest, and it is probable that further substantial areas will be transferred to forest in the foreseeable future.

Parts of the developed world now characterised by expanding forests previously suffered from trends of deforestation similar to those in the developing world today. Rapid deforestation affected areas such as the French Alps and parts of the eastern half of the United States in the nineteenth century. Indeed, many contemporary observers voiced fears remarkably similar to those currently expressed about the tropical forest. Yet in these and other areas, a transition from shrinking to stable or expanding forest areas was soon to be effected. In the case of France, for example, forests covered only 14 per cent of the land area at the time of the Revolution in 1789, whereas their extent is now almost twice as great (Prieur 1987). In areas such as the Alps, forest expansion resulted both from spontaneous regeneration on abandoned land or on land with lessening grazing intensities, and from deliberate reforestation (Douguedroit 1981).

A similar doubling, from a lower base, has taken place in Britain, and in some other countries the increase has been considerably greater. In Denmark, for example, the forest area increased from 4 to over 11 per cent of the land area during the last 150 years. In the United States, expansion has perhaps been less impressive in percentage terms, but the transition was nevertheless abrupt and dramatic: Williams (1988) writes of the 'death and rebirth of the American forest'.

The factors underlying and determining the transition from shrinking to expanding forests in the developed world are as yet poorly understood. It therefore cannot be confidently predicted whether or when a similar transition will take place in the developing world. Population growth rates may well be one important variable, although they are unlikely to be the only one. The statistical problems previously indicated are one reason why it is difficult to analyse in detail the inverse relationship that appears to exist between trends of population and of forest area. Calculated values of correlation coefficients are typically modest: for example, Allen and Barnes (1985) report a correlation coefficient of −0.492 for African and Asian countries in the 1970s. Analysis of the relationship is complicated by factors such as population densities, degrees of urbanisation and extents of national territories.

In general, however, countries with high rates of population growth usually have high rates of deforestation, and those with stable populations are often characterised by expanding forests. One fundamental question concerns the role of political factors, and especially of policies relating to forestry in general and to afforestation in particular. One of the aims of this volume is to examine the introduction of such policies in a variety of countries of contrasting geographical settings and political and historical backgrounds.

The adoption of afforestation policies does not necessarily lead to a forest transition. Afforestation may proceed alongside deforestation: indeed, it has not been unusual for native forest to be removed to make way for plantation forest. Also afforestation may fail to keep pace with unintentional loss of forest. In Greece, for example, forest fires were reported in the 1980s as destroying 25,000–120,000 ha of forest annually while 're-afforestation' amounted to 3000–4000 ha (Modinos and Tsekouras 1987). In Thailand, despite recent increases in afforestation rates, deforestation proceeded at an annual rate of 2.5 per cent during the 1980s (WRI 1990). Similar trends are to be found in numerous other tropical countries.

Tables 1.1 and 1.2, indicating reported forest trends in the countries discussed in this volume, clearly imply that deforestation can accompany afforestation. Conversely, it is possible for a country's forest area to expand without the adoption of a formal afforestation policy. Marginal land abandoned by agriculture, for example, may undergo spontaneous reforestation by natural regeneration. Nevertheless, most countries now characterised by stable or expanding

**Table 1.2** Trends in areas of forest and woodland, 1965–89 (thousand ha)

| | 1965 | 1970 | 1975 | 1980 | 1985 | 1989 |
|---|---|---|---|---|---|---|
| UK* | 1,804 | 1,879 | 2,020 | 2,102 | 2,273 | 2,364 |
| Ireland | 194[1] | 245[e] | 298 | 317 | 330 | 341[e] |
| Denmark | 3,99[1] | 472 | 484[u] | 493 | 493 | 493[e] |
| Hungary | 1,422 | 1,471 | 1,545 | 1,610 | 1,648 | 1,688 |
| Spain | 11,616[1] | 11,500 | 14,944 | 15,270[e] | 15,614 | 15,650[e] |
| Algeria | 3,045[2] | 3,700[e] | 4,122 | 4,384 | 4,384[e] | 4,699[e] |
| China | 109,180[e] | 111,524[e] | 142,224[e] | 136,365[e] | 130,865[e] | 126,465[e] |
| Chile | 20,686[1] | 15,420[u] | 8,680[e] | 8,680[e] | 8,800 | 8,800[e] |
| New Zealand | 6,232[3] | 6,280 | 7,095[e] | 6,960[e] | 7,160[e] | 7,320[e] |
| Australia | | 133,700[u] | 133,700[u] | 105,884 | 106,000 | 106,000 |
| Japan | 25,402 | 25,043 | 25,030[e] | 25,198 | 25,105 | 25,105 |
| USA[4] | 302,049 | 292,100[e] | 290,642 | 297,100[e] | 295,500[e] | 293,900[e] |

* These figures include Northern Ireland
e estimated figure u unofficial figure
1 Based on data from World Forest Inventory 1958 or 1963
2 For 1961
3 Excluding fern scrub, second growth and barren lands.
4 Excluding Alaska and Hawaii. Also excluding forest areas reserved in parks and for other uses.

*Note*: that estimates and other quoted figures may be adjusted in subsequent volumes.
*Source*: *FAO Production Yearbook*, 1966, 1971, 1976, 1981, 1986 and 1990.

forest areas have adopted policies of afforestation. On the other hand, some countries with afforestation policies that are ambitious in terms of area are still suffering deforestation or at best have static forest extents.

## National afforestation experiences

The countries discussed in subsequent chapters have been selected to represent a variety of settings and experiences. Although afforestation is spreading to many tropical countries, the emphasis in this volume is on countries with longer histories of forest expansion.

The first group of five European countries – Britain, Ireland, Denmark, Hungary and Spain – have relatively long-established policies of afforestation and substantial areas of afforestation. In some of these countries the native forest has almost completely disappeared, and most of the present forest area is in the form of plantations. The seven countries outside Europe – Algeria, China, Chile, New Zealand, Australia, Japan and the United States – are obviously geographically and politically diverse, and offer a wide range of forest histories and afforestation experiences.

Britain and Ireland, though environmentally similar, offer interesting contrasts in terms of politics and of policies towards land and forests. Both are characterised by a major expansion of the forest area in the twentieth century, but the rate and pattern of progress have differed between the two countries, as have the instruments of afforestation policies (Chapters 2 and 3). Afforestation began earlier in Denmark (Chapter 4), but the significance of historical and political factors, already emerging as a theme in the comparison of Britain and Ireland, is again apparent. A different variant of the same theme is clear also in the case of Hungary (Chapter 5). That country suffered major boundary changes and the loss of much of its well-forested territory after World War I, and thereafter initiated afforestation policies. The theme of *type* of afforestation, which has become the focus of major controversy in parts of Europe and increasingly also in other parts of the world, is developed in the chapter on Spain (Chapter 6).

The next two chapters, on Algeria and China, outline afforestation experiences that are unique and apparently spectacular. In both cases, the published statistics appear to indicate major achievements in afforestation. Both countries have suffered from severe deforestation (in the case of China over a very long period) and from serious environmental problems resulting from it (e.g. Murphey 1983; Thirgood 1986). Both have extensive tracts of land with severe climatic constraints which make the restoration of forest difficult. Both have suffered from major political upheavals. In Chapters 7 and 8, the background to their afforestation programmes is outlined, and the actual achievements evaluated.

The next three chapters, on Chile, New Zealand and Australia, review the origins and progress of afforestation in these countries. Afforestation generally began later than in European countries, and overlapped in time with forest clearance to a much greater extent than in most of Europe. Major industrial plantations have been established, especially in Chile and New Zealand, and exports of forest products have increased rapidly in volume and value in recent years. The afforestation programmes have demonstrated a degree of volatility, not least in accordance with the prevailing political climate. In the chapter on Chile, the social and environmental effects of such programmes are critically reviewed, while in that on New Zealand the increasing integration of the country's forest plantations into international systems of ownership and control is highlighted. In the case of Australia, two sectors are distinguished – afforestation for wood production, and tree planting for environmental reasons.

Japan was perhaps the first country to experience a forest transition. As early as the seventeenth century it was responding to the problem of shrinking forests and increasingly scarce forest products by introducing measures of forest conservation and reforestation (e.g. Osaka 1983; Totman 1982; 1984; 1986). Its forests came under renewed pressures in the present century, especially during World War II and the Korean War. Renewed activity in afforestation and reforestation followed (Chapter 12). Finally, the case of the United States of America is considered in Chapter 13. The American case is indeed dramatic (Williams 1989b). When, after many decades of deforestation, predictions of a timber famine were being made at the beginning of this century, conditions were already changing. The driving force of forest clearance for agriculture was largely spent, and abandoned land was already beginning to revert back to forest. Ironically the first manifestation of an American afforestation policy – on the prairie lands – fizzled out just before large-scale reversion to forest began elsewhere in the country. Subsequently, clearly defined afforestation policies of the type found in

**Table 1.3** Recent reforestation rates

| | (thousands or millions (M) of hectares per year) | |
|---|---|---|
| | UNEP | WRI |
| China | 4.6M | 4.6M |
| USSR | 4.5M | 4.5M |
| USA | 1.8M | 1.8M |
| Brazil | 450 | 449 |
| Japan | 240 | 240 |
| India | 138 | 173 |
| Turkey | 82 | 82 |
| Chile | 74 | 93 |
| Spain | 92 | 92 |
| Algeria | 52 | 66 |
| Australia | 62 | 62 |
| German FR | 62 | 62 |
| New Zealand | 43 | 43 |
| Britain | 40 | 40 |
| Hungary | 19 | 19 |
| Greece | 19 | 19 |
| Ireland | 9 | 9 |

*Note*: UNEP and WRI figures are averages for reforestation during the 1980s, and refer to the establishment of plantations for industrial and non-industrial uses. According to these authorities, some countries report regeneration as afforestation.
*Sources*: compiled from UNEP (1991) and WRI (1990).

some European countries have failed to materialise, yet around 1 Mha of forest is now planted annually. The United States clearly serves as a reminder that forest trends are not determined by forest policy alone, and that they may be profoundly affected by agricultural trends and policies.

Table 1.3 indicates the approximate scale of 'reforestation' in the countries examined in this volume and in a selection of other lands. It is included to provide a point of reference, context and perspective for the ensuing chapters, in which more precise estimates of afforestation rates are made. The table should be viewed with caution, for a number of reasons. One is the statistical problem that has already been mentioned. In addition to the lack of standardised definitions and reporting systems, it should be noted that targets and actual achievements are not always carefully distinguished. Survival rates are variable. There may also be considerable volatility in planting rates within individual countries, even if general policies of forest expansion are maintained. Finally, as will become clear in the ensuing pages, it should be emphasised that the effectiveness of afforestation policies cannot be measured solely in terms of planting rates or extents.

## References

Allen, J.C. and Barnes, D.F. (1985) The causes of deforestation in developing countries, *Annals, Association of American Geographers* 75: 163–84.

Braden, K. (1991) 'Managing Soviet forest resources', in P.E. Pryde, *Environmental management in the Soviet Union*. Cambridge University Press, Cambridge and New York, pp 112–34.

Douguedroit, A. (1981) Reafforestation in the French Southern Alps, *Mountain Research and Development* 1: 245–52.

Evans, J. (1986) Plantation forestry in the tropics – trends and prospects, *International Tree Crops Journal* 4: 3–15.

Evans, J. (1992) *Plantation forestry in the tropics* (second edition) (first edition, 1982). Clarendon, Oxford.

FAO (1967) World symposium on man-made forests, *Unasylva* 21 (86–87).

FAO (1988) Forestry policies in Europe, *FAO Forestry Paper 86*, FAO Rome.

FAO (annual) *Production yearbook*. FAO, Rome.

Glacken, C.J. (1967) *Traces on the Rhodian shore*. University of California Press, Berkeley.

Grainger, A. (1986) Deforestation and progress in afforestation in Africa, *International Tree Crops Journal* 4: 33–48.

Hughes, J.D. with Thirgood, J.V. (1982) Deforestation, erosion and forest management in ancient Greece and Rome, *Journal of Forest History* 26: 60–75.

Kunstadter, P. (1990) Impacts of economic development and population change on Thailand's forests, *Resource Management and Optimization* 7: 171–90.

Lanly, J.-P. (1982) Tropical forest resources, *FAO Forestry Paper 30*, FAO, Rome.

Logan, W.E.M. (1967) FAO world symposium on man-made forests and their industrial importance, *Unasylva* 21(86–7): 8–23.

McDonald, S.E. and Krugman, S.L. (1986) Worldwide planting of Southern pines, *Journal of Forestry* 84: 21–4.

Mather, A.S. (1990) *Global forest resources*. Belhaven, London.

Modinos, M. and Tsekouras, G. (1987) Forestry: Greece, *European Environmental Handbook*. DocTer, London, p 262.

Murphey, R. (1983) 'Deforestation in modern China', in R.P. Tucker and J.F. Richards (eds.) *Global deforestation and the nineteenth century world economy*. Duke University Press, Durham NC, pp 111–28.

OECD (1991) *Environmental data compendium*. OECD, Paris.

Osaka, M.M. (1983) 'Forest preservation in Tokugawa Japan', in R.P. Tucker and J.F. Richards (eds.) *Global deforestation and the nineteenth century world economy*. Duke University Press, Durham NC, pp 129-45.

Persson, R. (1974) World forest resources: a review of the world's forest resources in the early 1970s, *Research Note 17*, Royal College of Forestry, Stockholm.

Postel, S. and Heise, L. (1988) Reforesting the earth, *Worldwatch Paper 83*, Washington.

Prieur, M. (1987) Forestry: France, *European Environmental Yearbook*. DocTer, London, pp 252–8.

Richardson, S.D. (1990) *Forests and forestry in China*. Island Press, Washington.

Savill, P.S. and Evans, J. (1986) *Plantation silviculture in temperate regions with special reference to the British Isles*. Clarendon, Oxford.

Sedjo, R.A. (1984) An economic assessment of industrial forest plantations, *Forest Ecol-*

*ogy and Management* 9: 245–57.

Sedjo, R.A. (1991) 'Forest resources: resilient and serviceable', in K.D. Frederick and R.A. Sedjo (eds.) *America's renewable resources: historical trends and current challenges*. Resources for the Future, Washington, pp 81–123.

Thirgood, J.V. (1986) The Barbary forests and forest lands, environmental destruction and the vicissitudes of history, *Journal of World Forest Resource Management* 2: 137–84.

Totman, C. (1982) Forestry in early modern Japan 1650-1850, *Agricultural History* 56: 415–26.

Totman, C. (1984) 'From exploitation to plantation forestry in early modern Japan', in H.K. Steen (ed.) *History of sustained-yield forestry: a symposium*. Forest History Society, Santa Cruz, pp 270–80.

Totman, C. (1986) Plantation forestry in early modern Japan: economic aspects of its emergence, *Agricultural History* 60: 23–51.

Turner, J. and Gessel, S.P (1990) 'Forest productivity in the Southern hemisphere with particular emphasis on managed forests', in S.P. Gessel, D.S. Lacate, G.F. Weetman and R.F. Powers (eds.) *Sustained productivity of forest soils*, Proceedings of 7th North American Forest Soils Conference, University of British Columbia, Vancouver, pp 23–39.

UNEP (United Nations Environment Programme) (1991) *Environmental data report* (third edition 1991/92). Blackwell, Oxford.

Williams, M. (1988) 'The death and rebirth of the American forest: clearing and reversion in the United States 1900–1980', in J.F. Richards and R.P. Tucker (eds.) *World deforestation in the twentieth century*. Duke University Press, Durham NC, pp 212–29.

Williams, M. (1989a) Deforestation: past and present, *Progress in Human Geography* 13: 176–208.

Williams, M. (1989b) *Americans and their forest: an historical geography*. Cambridge University Press, Cambridge and New York.

WRI (World Resources Institute) (1989) *World resources 1988–89*, Basic Books, New York.

WRI (1990) *World resources 1990–91*, Oxford University Press, Oxford and New York.

# 2 Afforestation in Britain

*Alexander Mather*

## Introduction

The most striking change in rural land use and landscape in Britain in the twentieth century has been the expansion of forests. At the beginning of the century, only 5 per cent of the land surface was under forest. Today it is 10 per cent. In some parts of Britain, the increase has been much greater. In Scotland, for example, forests now occupy 14.5 per cent of the land area, compared with 4.5 per cent in 1905. Planting rates have fluctuated over the century, and patterns of afforestation have varied. Nevertheless, the afforestation effort has been sustained over seven decades, despite major changes in the objectives of afforestation policy and in the economic, social and political climates in which it has been set. This continuity of effort, despite changing circumstances and despite opposition on both economic and environmental grounds, is one of the main features of afforestation in Britain.

In recent years opposition to the forms of commercial coniferous afforestation that have characterised much of the present century has been strong, and planting rates have fallen. We are at present in one of the periodical interludes of questioning of afforestation policy. On the basis of past experience, it is to be expected that forest expansion will continue in the last few years of the present century and perhaps well into the next. Nevertheless, its form and pattern may well differ from previous decades.

## Afforestation: organisation and structure

Forest is the natural vegetation of most of Britain. Mixed oakwoods at one time occupied much of the lowland area, with pine-dominated forest in part of the Scottish Highlands and birch in the far north. Most of this natural woodland has long since been removed: one of the characteristics of forest and woodland in Britain is the extent to which it is planted rather than natural.

Today, Britain is poorly wooded compared with most of its European neighbours, but the type and extent of its forest and woodland vary greatly. In England more than half of the woodland area consists of broadleaves, in some cases long established and often planted and managed for sport or amenity rather than primarily for wood production. In Scotland, more than 90 per cent of the forest area is coniferous, and most of it is managed primarily for wood production.

Forest ownership in Britain is divided between the state and private sectors in the approximate ratio of 45:55 (Table 2.1). Prior to 1919 when the state forestry

**Table 2.1** Forest area (thousands of hectares) *c.* 1990

| | Scotland | England & Wales | Britain |
|---|---|---|---|
| Forestry Commission | 526 | 372 | 898 |
| Private | 520 | 686 | 1206 |
| Total | 1,046 | 1,058 | 2,104 |

These figures relate to high forest. In addition, there are further areas of coppice and unproductive woodland: the total area of forest and woodland is approximately 2.4 million ha.
*Sources*: *Annual Abstracts of Statistics; Scottish Abstract of Statistics; Forestry Commission Facts and Figures.*

service (the Forestry Commission) was established, it lay almost wholly in the private sector. Most of the afforestation carried out since then was undertaken by the Forestry Commission, although the private sector has been responsible for much of the effort since the early 1980s. Most of the planting in the private sector, however, has been grant-aided by the Forestry Commission, in its role as forestry authority. The Forestry Commission has had two main roles: in its 'forestry enterprise' capacity, it acquired, planted and managed forest land, while in its 'forestry authority' role it administered instruments of policy such as planting grants and felling licences. These two roles were formally separated in 1992.

From the early 1980s, the role of the Forestry Commission in carrying out afforestation began to decrease, and a policy of partial privatisation of its estate was introduced. By 1989, around 120,000 ha of FC forest had been transferred to the private sector, amounting to about 10 per cent of the FC area. (In that year further disposals of around 100,000 ha were proposed, to be achieved by the end of the century.) Little information is published on the private purchasers, but analysis of the early disposals indicated that the nature of the purchaser was likely to depend on the type and extent of property offered for sale (Mather and Murray 1986). Small, outlying plantations were likely to be acquired by local farmers, while larger forests better located in relation to potential markets tended to be purchased by corporate investors such as pension funds.

Few published data exist on the nature of private forest ownership in Britain, but some characteristics are clear. Firstly, most private ownership prior to the 1960s was in the hands of large landed estates, often with long forestry traditions. Secondly, non-traditional private investors became prominent between the mid-1960s and late 1980s: they are estimated to have accounted for around 35 per cent of the private forest ownership in Scotland in the mid-1980s, compared with 43 per cent in the hands of the traditional estates (Mather 1987). (Traditional estates own a much higher proportion of private woodland in England, and 'private investors' a correspondingly smaller share.) Thirdly, corporate ownership is poorly developed. A number of financial institutions such as pension funds and insurance companies invested in forest land and in some cases carried out afforestation during the 1970s and 1980s in particular, and were estimated to own around 10 per cent of forest land in Scotland in the mid-1980s.

Ownership – and afforestation – by industrial companies is rare in Britain, compared with some other countries. Farmers and small land owners have also been poorly represented amongst forest owners and afforesters in Britain. Private forest ownership and, until very recently, private afforestation have been mainly associated either with large, long-established estates or with private investors in the form of high-income, urban-based individuals, including sports stars and entertainment celebrities.

## Historical background

Compared with most European countries, as well as with countries in the wider world, Britain experienced early and severe deforestation. Much of this deforestation resulted from clearance of land for agriculture, from prehistoric times onwards. By the medieval period, much of the country was largely denuded of its forest cover (see Anderson (1967) and James (1981) for histories of forestry in Scotland and England respectively). Deforestation was especially extensive in the lowlands. In lowland Scotland, for example, woods and forests probably occupied no more than 2–3 per cent of the land area in the mid-eighteenth century, when more reliable maps and records became available.

The disadvantages of treelessness became all too apparent from the Middle Ages onwards. The shortage of wood for naval shipbuilding was a particular problem that attracted the attention of the state. In both Scotland and in England attempts were made to protect the remaining forest resource and to carry out afforestation for this purpose. In the case of the latter, the old royal (hunting) forests were the main foci of activity. In the case of the former, legislation dating from the beginning of the sixteenth century required land-owners in areas where there was no significant forest to plant at least one acre (0.4 ha) (Caird 1980).

Such efforts were usually temporary and were often associated with periods of crisis (such as wars) which heightened the awareness of timber scarcities. Planting was usually confined to very small areas (for example, around country houses). By the late eighteenth and early nineteenth centuries, however, more extensive afforestation was being carried out, particularly by a number of private land owners in Scotland. By this time, more interest was being shown in conifers grown for timber, and mainland European species such as larch and Norway spruce were now being extensively planted on some estates. During the first half of the nineteenth century, North American species such as Sitka spruce and Douglas fir were introduced.

Planting during the late eighteenth and early nineteenth centuries was set in the context of agricultural change and the enclosure of land. In some parts of Britain enormous lengths of hedges and hedgerows were established, and tree planting was an integral part of the landscaping that was carried out by many estate owners around their country houses. In a few localities, more extensive commercial plantations were established, oriented towards timber demand for shipbuilding and other industrial purposes.

While there was undoubtedly an awakening of interest in forestry and silviculture around this time, the afforestation effort was not long sustained. By the middle of the nineteenth century it had faded, in the face of cheap timber imports from Canada and elsewhere. Such planting as was carried out during the second half of the century was motivated mainly by considerations of amenity and field sports. Furthermore, the nature of demand for timber was radically changing. The use of iron and steel for shipbuilding meant that there was little incentive to continue to manage the existing oakwoods. In short, forestry was at a low ebb, at least in terms of expansion and management of the existing forest areas. It seemed that the beginning of the nineteenth century had turned out to be a false dawn. It transpired, however, that the second half of the nineteenth century was a period of darkness before the real dawn.

## The dawning of the age of afforestation

The second decade of the twentieth century saw the beginning of significant and sustained afforestation in Britain. The crisis of World War I was the trigger that set in motion the drive towards forest expansion. But while the war may have been the immediate cause, it is clear (with hindsight) that the political climate had been growing steadily more favourable for state-assisted afforestation during the preceding decades. Actual achievements in terms of areas planted during that period were negligible, but the shortcomings of *laissez-faire* policies towards forestry were being increasingly realised and the first signs of government intervention were beginning to appear.

A Parliamentary Select Committee on Forestry reported in 1887, and advocated that a Board of Forestry should be established and that it should have as one of its functions the provision of forestry education (Select Committee 1887). Ironically, instruction in forestry had become available a few years earlier, but only for students intending to practise forestry in India (James 1981). Little action ensued, but a Departmental Committee was set up by the Board of Agriculture in 1902 to consider British forestry. In addition to advocating expansion of education, it recommended that substantial demonstration areas be established and that local authorities should be encouraged to plant trees in water-catchment areas (Departmental Committee 1902).

The gradual build-up of momentum was further reflected in the report of the Royal Commission on Coast Erosion and Afforestation, published in 1909 (Royal Commission 1909). Originally set up in 1906 to examine coastal erosion, the Commission was in 1908 instructed also to consider afforestation, especially in relation to providing employment in periods of recession. Its report heralded (if not directly or immediately) a new age and scale of afforestation: two alternative schemes for state afforestation were proposed, one involving the annual planting of 150,000 acres (61,000 ha) over 60 years and the other 75,000 acres (30,000 ha) over 80 years. The former would have added 9 million acres to the then-existing 3 million acres of woodland in Britain (including at that time Ireland), the latter 6 million acres.

The proposed trebling or quadrupling of the existing woodland area clearly reflects a new scale of thinking and ambition, and it is perhaps not surprising that public and Parliamentary opinion was not immediately ready to translate such proposals into action. Further committees were subsequently appointed, but there was little concrete achievement by the time of the outbreak of war in 1914.

## The Acland Committee and the beginning of state afforestation

While the basis of a forestry infrastructure had partially and tentatively been laid over the previous few decades, it took the crisis of World War I to translate talk about afforestation into action. During the war, the disadvantages of relying on imported timber soon became all too apparent. As a result, the Forestry Sub-Committee of the Reconstruction Committee was set up in 1916. Usually known as the Acland Committee (after its chairman), it was composed of large land owners with experience in forestry, and representatives of various government departments. Its simple but far-reaching terms of reference were: 'To consider and report upon the best means of conserving and developing the woodland and forestry resources of the United Kingdom, having regard to the experience gained during the war' (Acland Committee 1917–18, p. 3).

The findings of the report made sorry reading. Britain was, with one exception, the least wooded country in Europe. Much of its woodland area was neglected, to the extent that production was only one-third of what it might have been under efficient management such as that practised in parts of mainland Europe. It produced only 8 per cent of its wood requirements, the remainder being imported. Imports had increased five-fold over the previous 70 years to meet the rapidly increasing demand for wood. This increase resulted not just from a growing population, but also from greatly increased per capita consumption as industrial development ran its course. Most of the growth in demand was for softwoods (especially for pitprops for the coal-mining industry, which had expanded very rapidly in the last quarter of the nineteenth century).

Whilst reliance on imports was a particular problem in time of war, the Acland Committee was convinced that the problem would not disappear at the end of hostilities. It noted that a decreasing share of imports was coming from the British Empire, and that most of them were from virgin forests that were liable to exhaustion. In other words, timber shortages on the world market were feared.

Some of the members of the Acland Committee may have been unlikely advocates of state intervention and especially of state acquisition of land, yet the clear recommendation was that a state forest authority, empowered to acquire and plant land, should be established forthwith. One of the large land owners on the committee – Lord Lovat – appended a note of reservation. In it he did not disagree with the recommendation of the committee, but considered it to be insufficiently strong and forceful in expression.

As well as identifying the immediate and longer-term problems of reliance on timber imports, the Acland Committee also considered that social benefits would accrue through the employment provided from a policy of afforestation, and that large areas of poor-quality land could be afforested without significant loss of food production.

A specific goal of rendering the country independent of imported timber for emergency periods of three years was suggested. This, it was estimated, would require the afforestation of 1.77 million acres (0.72 Mha) (representing an increase of about 60 per cent on the then-existing woodland area in the UK). This, it was proposed, should be carried out over a period of 80 years, with two-thirds of the planting being concentrated in the first 40 years. Most of the planting would be by the state, but grants would be provided to encourage private individuals and public bodies also to become involved in planting and in rehabilitating neglected woodlands. Various other considerations were also reviewed: for example, it was decided that the optimal spatial pattern, not least for reasons of security, should be one of wide dispersal rather than concentration. Perhaps the most significant conclusion, however, was that most of the afforestation should be coniferous. Most of the demand for pitprops and other industrial purposes was for softwood, whilst demand for hardwoods, with the disappearance of the traditional naval markets, had risen much less steeply. Indeed, some private owners had already converted their broadleaved woods to conifers.

The recommendations of the Acland Committee were soon translated into action, and funding was promised for the first ten years of the proposed programme. A state forestry authority – the Forestry Commission – was established in 1919, with afforestation at the core of its responsibilities. The primary policy objective was concerned with defence strategy: to permit self-sufficiency over limited periods of up to three years. Other objectives were related to insurance against a world-wide shortage of timber, and to the provision of rural employment in the hope of stemming rural depopulation.

## Progress between the wars

Progress in afforestation is illustrated in Figure 2.1. While still in its infancy, the Forestry Commission came under serious attack. In 1922, the Committee on National Expenditure (Geddes Committee) recommended its abolition, at a time of financial crisis in the aftermath of the war. The government rejected the recommendation, but funding was reduced and hence the planting programme was cut back compared with the Acland programme. At around the same time, however, the FC was given short-term funds to provide jobs, in a period of high unemployment, and in exasperation it referred to the 'grotesque' coincidence of cuts and additional funding (FC 3rd Annual Report 1922, p 3). The provision of employment was to prove a continuing theme during the inter-war period. By the 1930s, the shortfall in planting in the first ten-year state programme had been compensated by increased funding in an effort to relieve unemployment, especially around severely depressed industrial areas such as South Wales and North-east England.

Part of this 'social' thrust of afforestation involved the creation of forest-worker holdings (FWHs). These were intended as small-scale agricultural holdings, whose working could be combined with forest employment. During the early part of the century the idea of land settlement attracted much attention and some government support. The Acland Committee had itself advocated FWHs, but their creation was not an integral part of the initial afforestation programme and was not begun until 1924. By 1934, over 1200 had been created, and in some cases were occupied by unemployed coal-miners. The Forestry Commission considered that the scheme had been successful (FC 15th Report 1934). One problem was the difficulty of securing suitable areas of land, combining suitable location (especially in relation to centres of unemployment), land appropriate for FWHs and land capable of afforestation.

From the outset, it had been assumed that arable land would not be afforested.

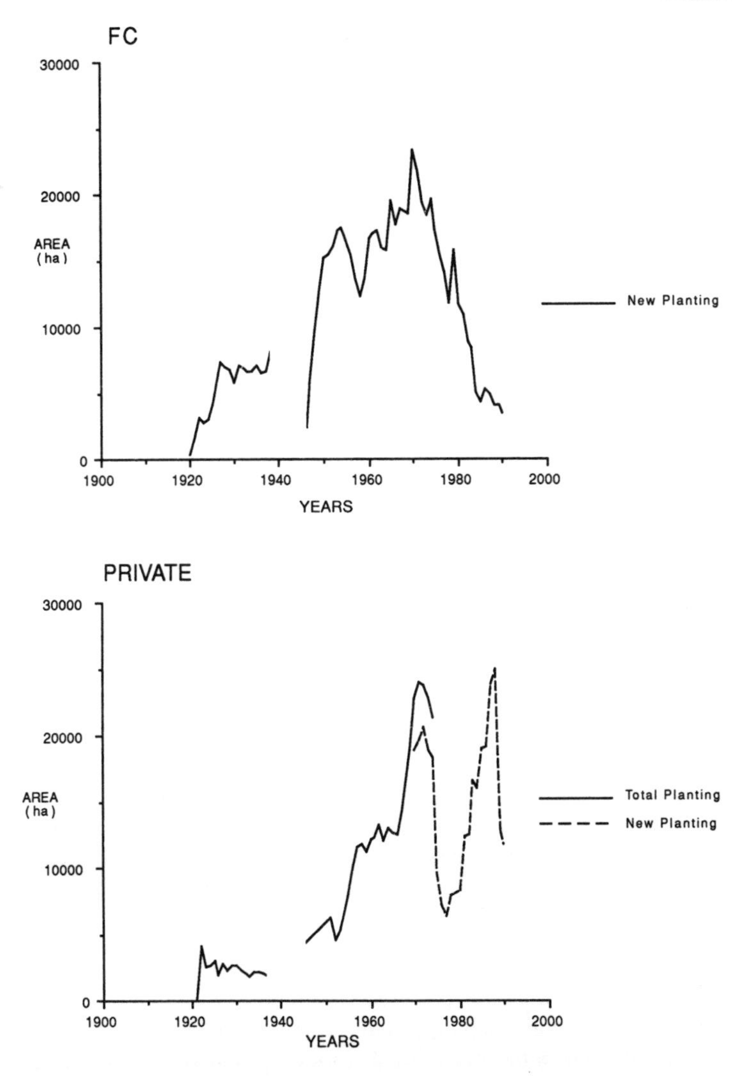


**Figure 2.1** State and private afforestation rates in Britain.

Planting was therefore restricted to 'waste' land and hill grazings. Initially, sizeable areas of lowland heath of low agricultural value (and including a number of areas of coastal sand dunes) were planted, as well as properties in the Highlands and uplands. It was not always easy to obtain land without extensive 'surplus assets' on the one hand (e.g. buildings, arable land) and without extensive unplantable areas (e.g. severely exposed or rocky land) on the other. By the mid-1930s, however, the FC was perceiving a new potential. Having considered how best to make use of the 'surplus and unplantable land' in a group of forests in Argyll in western Scotland, it decided that 'National Forest Park' status was appropriate. Thereafter recreational facilities such as camp sites and car parks began to be provided in areas accorded this status. In practice, a recreational function was being grafted on to the primary objectives set out by Acland: the first portent of this development was the admission to state forests of members of organised groups for recreational purposes in 1931.

The mid-1930s also saw the emergence of an issue that was to dog afforestation over the next half century. The first reference to conflicts between afforestation and preservation of amenity appeared. Following discussions with the Council for the Preservation of Rural England, the FC agreed not to acquire land for afforestation within a 300 square-mile area of the Lake District in northern England (FC 17th Report 1936).

## *Post-war forest policy* and post-war afforestation

The 1930s scene of accelerating planting rates, a growing social role of afforestation and the emergence of recreation and amenity issues was brought to an abrupt halt by the outbreak of World War II in 1939. The experience of World War I was repeated: timber shortages were encountered, the inter-war plantings were not yet mature, and the depredations in the existing woodlands during World War I had not yet been made good.

The outcome was the publication of *Post-war forest policy* in 1943 – the World War II equivalent of the Acland Report (Forestry Commissioners 1943). It was now estimated that 5 million acres (2 Mha) of 'effective' forest were required to meet the 'defence strategy' objective. To achieve this area, the afforestation of 3 million acres would be required, with existing woodlands making up the balance of 2 million acres. It was assumed that the afforestation would have to be carried out by the state, little having been achieved by the private sector during the inter-war period. Replanting of existing woodlands, however, would be encouraged in the private sector by a new Dedication Scheme, in which land owners would 'dedicate' land to forestry in return for grant aid. Some thought was again given to the optimal spatial pattern. Local and general markets were both identified. While the former could be supplied by a large number of small, dispersed woods, it was considered that the latter were best served from large forest units. Some incipient 'forest regions' (such as around the Moray Firth and in the Borders) had already begun to emerge, and it was recommended that the possibility of creating such areas be given due attention in land acquisition. The scene seemed to be set for a new, reinvigorated era of afforestation, with planting rates almost double their inter-war levels.

A cautionary note was sounded at the outset by government. At the end of the war, the Minister of Agriculture observed that the proposals were 'large', and that they would require careful consideration in relation to their effect on agriculture. Accordingly, the proposals of *Post-war forest policy* were not accepted in full, but funding was provided for the first five years of the programme.

### *Progress in the state sector*

During the inter-war period, agriculture had been in a depressed condition, and the acquisition of hill-grazing land for afforestation had posed no major problems. During World War II, however, the dangers of over-dependence on imported food became all too apparent, and a major programme of agricultural expansion began (and was to continue for four decades). The consequence for the afforestation programme was two-fold: land acquisition became more difficult overall, and such land as was made available for afforestation was that of the lowest agricultural quality. In effect, afforestation (at least by the FC) was relegated to the poorest hill grazings. Not only did the planting achievement soon begin to lag behind the rate implicit in the war-time target, but also the afforestation effort began to shift 'up the hill' onto poorer land, with both locational and silvicultural consequences. A shift towards the north and west soon became apparent, and restriction to the poorest areas of land restricted the range of tree species that could easily be used.

Although planting rates did increase rapidly in the late 1940s (Figure 2.1) they soon stagnated, mainly as a result of problems of land acquisition. Indeed, it seemed by the mid-1950s that the momentum of the post-war planting programme had been lost. A further blow was the recognition that the long-established policy objective of maintaining a standing reserve of timber for use during a period of war was now largely irrelevent. In 1957, it was wryly observed that 'if a future war were to be fought with nuclear weapons ... it is very doubtful if the time factor would permit reserves of standing timber to aid us in the prosecution of hostilities' (Zuckerman Committee 1957). It seemed that the *raison d'être* of afforestation had been removed at a stroke.

Yet the afforestation programme did not grind to a halt. After a period of declining activity in the mid-1950s, it accelerated steadily until the early 1970s (Figure 2.1). Objectives now became social and commercial, rather than strategic. The production of wood as a raw material for industry (as opposed to the maintenance of a standing reserve of timber) was now emphasised as a policy objective, along with profitable management of the forest estate and the provision of employment in depopulating rural areas (e.g. FC 44th Annual Report 1963).

In the early years of the post-war programme, social factors and the provision of employment played little direct part in forestry policy. By the mid-1950s, however, a small regional element was introduced in the form of a special planting programme in crofting areas in north-west Scotland, and from the late 1950s (and the demise of the 'strategic reserve' objective) social and regional issues assumed roles of increasing importance. In a 1958 statement on forestry policy, for example, it was stated that 'in deciding where planting shall take place, special attention will be paid to upland areas, particularly in Scotland and Wales, where expansion of forestry will provide needed diversification of employment and important social benefits' (FC 39th Annual Report 1957–8). Whereas the earlier intention had been to split new planting on an equal basis between Scotland on the one hand and England and Wales on the other, Scotland was now emerging as the main area of afforestation in Britain (Table 2.2).

Social benefits through employment provision had been an underlying (and fluctuating) theme since the time of the Acland Committee. At various stages, the provision of FWHs and of opportunities for unemployed coal-miners and industrial workers had been emphasised. Throughout, it had been asserted that afforestation would provide far more jobs per unit area than alternative land uses such

**Table 2.2** Scotland's share (%) of new planting in Britain

| | FC | Private |
|---|---|---|
| 1950 | 48 | na |
| 1960 | 58 | na |
| 1971 | 81 | 74 |
| 1980 | 90 | 83 |
| 1990 | 99 | 80 |

(New planting is not distinguished from restocking in private sector prior to 1971)
*Source*: compiled from data in FC Annual Reports.

as hill farming, and forestry villages had been seen as a means of repopulating some remote rural areas (Wonders 1990). Under the Labour governments of the 1960s, planting targets for Scotland were increased in three successive years, with a view to providing employment opportunities in the crofting counties of north-west Scotland in particular. The relative importance of Scotland increased further: conflicts between afforestation and landscape conservation in the national parks of England and Wales further aggravated problems of land acquisition there, and the target for England and Wales was not raised. By the late 1960s, significant land acquisition in England was largely confined to the northernmost counties of Northumberland and Cumberland.

The effects of these decisions to increase planting rates were soon apparent. By 1970, state afforestation rates were at a record level, in excess of 50,000 acres (20,000 ha) per year. With the continuing policy of protecting all but the poorest agricultural land, however, much of this area was of modest productivity and hence of modest economic performance under forestry.

### *Progress in the private sector*

Experience during the inter-war years had not encouraged much optimism about the rate of new planting that might be expected in the private sector. Hence it was assumed in *Post-war forest policy* that activity in that sector would be confined mainly to replanting and rehabilitating existing woodlands. By the mid-1950s, however, total planting (i.e. including replanting) in the private sector was exceeding the target, and rates were accelerating (FC 37th Annual Report 1956). Statistics for new planting (i.e. afforestation) and replanting were not distinguished in FC reports at the time, but it is clear that rates of new planting were increasing, and indeed that they 'took off' during the late 1960s (Figure 2.1).

Initially, most planting (both new planting and restocking) was carried out by traditional estate owners, but by the early 1960s private investors (other than traditional land owners) were being increasingly attracted. Planting grants under the Dedication Scheme were one incentive: the other was favourable treatment of investment in forestry under both income and capital-tax arrangements. The emergence of private financial syndicates on the forestry scene was first recorded in FC annual reports in 1958; in some cases they were assembled by firms of accountants and motivated primarily by tax-saving. By the mid-1960s an awareness of the tax advantages of forestry investment was widespread, and in the latter part of the decade new planting rates in the private sector rose to 20,000 ha per year (Figure 2.1) – matching those in the state sector, and resulting in record rates of total afforestation in excess of 40,000 ha per year.

Whereas the FC had to seek the 'clearance' of land from the agricultural ministry prior to afforestation, no such constraint existed in the private sector. Market forces dictated the pattern of land transfer. Some of the best conditions for afforestation (including long growing seasons, high rainfall and easily planted land) existed in south-west Scotland, and huge areas of land were planted there in particular. Forestry management companies emerged to acquire, plant and manage land on behalf of their 'private investor' clients, many of whom had no tradition of ownership of land or forests.

## Reaction sets in: downturn in the 1970s

The record rates of afforestation were not long sustained. With the advent of a Conservative government in 1970, a policy review was undertaken, and the result was a cutting back in the state planting target to a *combined* figure (i.e new planting and restocking) of 22,000 ha per year. A Treasury cost–benefit review had cast serious doubt on the financial performance of forest investment, but forestry interest-groups mounted a vigorous defence campaign and a programme of afforestation continued. Ambivalence in policy was reflected in the statement that while social factors would continue to be important elements in policy, they would not be pursued regardless of cost (MAFF 1972).

The early 1970s also marked the end of a chapter in the private sector. In the face of the disquiet of farmers about the rapid and uncontrolled 'loss' of land into forestry, new procedures were introduced to regulate transfer. Whilst planting grants would continue to be available under a new Dedication Scheme, applications would be approved only after consultation with some or all of a number of bodies, including the agricultural ministries, the local planning authorities, and conservation agencies. In short, an influence of planning was now being brought to bear on private-sector planting.

Soon after these new institutional arrangements were introduced, however, private planting rates fell drastically. There had been some loss of confidence arising from the uncertainty that had set in during the review period, and the replacement in 1975 of Estate Duty by Capital Transfer Tax by the new Labour government presaged a collapse. Whereas the regulations under the former were very favourable to forestry, those under the latter were not (at least at first). Private planting rates declined to one-third of their previous level (Figure 2.1).

The second half of the 1970s was a period of uncertainty in both sectors. By now the target of an 'effective' forest estate of 2 million ha was in prospect (and indeed was achieved in 1983). Further programmes of afforestation were advocated by the FC (1977) and CAS (1980): amongst the arguments advanced in favour of further expansion were the possibility of future wood shortages and the availability of suitable land of low agricultural value.

The new Conservative government issued a policy statement at the end of 1980: afforestation would continue at roughly the same rate as the average for the previous 25 years (subsequently specified as 30,000 ha per year). This expansion was held to be 'in the national interest', in terms of both reducing dependence on imported wood and providing employment. There was, however, a new feature of some importance: most of the new planting was to be carried out by the private sector, and the FC was instructed to embark on a programme of disposal of part of its estate (e.g. FC 61st Annual Report 1981). Planting rates in the state sector thereafter dwindled, whereas those in the private sector increased, under the twin incentives of a new planting grant (the Forestry Grant Scheme) introduced in place of the Dedication Scheme, and of tax benefits.

Investment in private forestry was actively promoted by a number of forestry management companies. Their clients were attracted by the opportunity of reducing tax bills and creating capital assets. The combination of planting grants and tax benefits accounted for around 70 per cent or more of the cost of afforestation, allowing the investor to create an asset at a fraction of its real cost. After a period of around ten years, when the tax advantages arising from expenditure on a plantation had been enjoyed, the private investor typically sold the forest, perhaps to a financial institution such as a pension fund. Official breakdowns of afforestation by type of investor are not available, but sample surveys in Scotland in the mid-1980s suggest that over 50 per cent of the planting was carried out by private investors, with the remainder divided approximately evenly between corporate investors on the one hand, and traditional estate owners and farmers on the other (Mather and Murray 1988).

The environmental effects of afforestation attracted increasing attention during the 1980s, especially when activity moved northwards to what came to be known as the Flow Country of the north of Scotland (e.g. NCC 1986). At the same time as these peatlands were being perceived increasingly positively in terms of afforestation, their value for purposes of nature conservation was being seen in a new light. A major conflict emerged between forestry and conservation interests, and the afforestation/conservation issue in the Flow Country became a *cause célèbre*. An alliance between environmentalists and those opposing the social injustice of 'subsidising' the rich through tax benefits brought the 1980s chapter of afforestation to an abrupt halt. In 1988, the income-tax advantages were discontinued. Planting grants were increased, through a new Woodland Grant Scheme, but the result was a sudden drop in planting rates reminiscent of that of the 1970s.

The forestry industry – or at least its afforestation arm – in recent years has been in the doldrums. Planting rates have stagnated, and the private forestry management companies, which a few years previously were so vigorous, have encountered hard times and restructuring. Once again, there is uncertainty about the future form and pattern of afforestation.

## Environmental issues and planning

A feature of the last twenty years, and in particular of the 1980s, has been the growth of disquiet about the environmental effects of afforestation. This disquiet has extended to both how and where planting is carried out, and to issues such as the effects of afforestation on streams and rivers, on wildlife habitat, and on landscape amenity.

Although it is true that some awareness existed prior to World War II about its possibly detrimental effects on landscape in areas such as the Lake District, afforestation was usually seen as positive or neutral in environmental terms. For example, some inter-war afforestation stabilised previously eroding areas of sand dunes. With this background, it is not surprising that when the foundation of the British planning system was laid in the Town and Country Planning Act of 1947, afforestation was excluded from control. In any case, it was intended then that afforestation would be carried out by the state, so that further state control was not deemed necessary.

By the time of the high planting rates of the late 1960s and 1970s, however, disquiet about the nature of forest expansion was being voiced. In 1974, 'consultation procedures' were introduced, whereby the Forestry Commission consulted with the local planning authorities, and with the Nature Conservancy Council and the agricultural ministry, over some categories of applications for planting grants. Following on these consultations, the FC could refuse the application, or impose certain conditions on approval. In other words, the planning authorities could now exert some influence, indirectly through the system of grant aid to private afforestation. (The applicant could still proceed to afforest if the application for grant aid were refused, and indeed would still be eligible for tax relief. In practice, however, very little planting has taken place without grant aid. Private forestry interests and the Forestry Commission had an informal agreement whereby the consultation procedures would not be by-passed.)

In reviews of the working of these consultation procedures, it has been found that in the majority of cases, the requests of the planning authorities are satisfied (e.g. Brotherton and Devall 1988; Heatherington 1988; Mather 1991a). But significant numbers of cases involving large afforestation schemes encounter problems and difficulties, and the consultation procedures were quite unable to contain (and far less to resolve) the conflicts between forestry and conservation interests in areas such as the Flow Country of northern Scotland. One of the main objections to the procedures was the 'case by case' approach: individual applications might be unobjectionable, but the cumulative effect of numerous afforestation schemes in a district might be the extensive loss of wildlife habitat or extensive landscape change. As a result, by the late 1980s interest was growing rapidly in 'indicative forest strategies', whereby an 'indicative' division was made at the regional or sub-regional scale into areas 'preferred', 'possible' or 'sensitive' in relation to afforestation. These strategies, worked out in conjunction with forestry and other interests, will be incorporated into the regional structure plans in due course (Mather 1991b).

It is perhaps significant that the introduction first of the consultation procedures and then of the indicative forestry strategies followed periods of high planting rates in the private sector. Ironically, both of these extensions of the influence of planning on afforestation were followed by large decreases in annual planting rates, resulting in turn from changes in tax regulations at the national level.

The growth of the influence of planning on afforestation has stopped short of

**Table 2.3** Scottish forest area by species and ownership

| | FC | Private | Total area (ha) |
|---|---|---|---|
| | (per cent) | | |
| Conifers | 99 | 80 | 765,633 |
| Scots pine | 12 | 24 | 144,371 |
| Lodgepole pine | 17 | 6 | 103,924 |
| Sitka spruce | 52 | 31 | 364,601 |
| Norway spruce | 7 | 6 | 54,707 |
| Jap/H larch | 7 | 6 | 52,146 |
| Broadleaves | 1 | 20 | 76,567 |
| Oak | 1 | 4 | 16,551 |
| Beech | <1 | 3 | 10,496 |
| Birch | <1 | 4 | 16,647 |

*Source*: *Scottish Abstract of Statistics* 1990 (based on Forestry Commission census 1980).

the introduction of full planning controls. These have frequently been advocated, not least by the Countryside Commission (e.g. 1987), but have been strenuously opposed by forestry interests. In the face of demands for stronger planning controls, however, both the Forestry Commission (as forestry authority) and Timber Growers UK (representing private woodland owners) have taken steps to make afforestation more environmentally friendly. The latter, for example, has encouraged its members not to undertake planting outside the consultation procedures. It also took the initiative in producing a code of practice which sought to curb adverse environmental impacts arising from practices such as ploughing and ditching prior to afforestation (TGUK 1985). The Forestry Commission has itself further developed this approach, through sets of guidelines relating to such practices. This approach, emphasising good practice but avoiding control or coercion, fitted well with the philosophy of 'voluntarism' that characterised many aspects of planning and environmental management in Britain in the 1980s.

Environmental issues are also reflected in the changing nature of afforestation grants. Under the Woodland Grant Scheme (WGS) introduced in 1988, much more attention is focused on environmental questions (such as the possible impact of planting on nearby watercourses) than in previous grant schemes. Timber production was the primary objective under previous schemes: other objectives such as nature conservation, sport and amenity may now have primacy.

The introduction of a Broadleaved Woodland Grant Scheme (BWGS) in 1985 was an earlier manifestation of the shift away from primacy of wood production as a policy objective. The shrinking legacy of broadleaved woodland, especially in the face of replanting with conifers, gave rise to much concern amongst environmental interests, particularly in England. This trend of coniferisation had been underway for many decades, but was popularly perceived as having reached serious proportions by the early 1980s (see Watkins 1986). A Forestry Commission census in 1980 indicated that around 99 per cent of the FC forests and 80 per cent of private woodlands were composed of conifers (Table 2.3). The BWGS

**Table 2.4** Composition of new planting, Scotland

| | FC | | | Private | | |
|---|---|---|---|---|---|---|
| | BL (ha) | Con (ha) | %BL | BL (ha) | Con (ha) | %BL |
| 1980 | 83 | 14,052 | 0.6 | 131 | 6,801 | 1.9 |
| 1985 | 42 | 4,735 | 0.9 | 127 | 14,008 | 0.9 |
| 1986 | 41 | 3,997 | 1.0 | 245 | 17,047 | 1.4 |
| 1987 | 221 | 4,845 | 4.6 | 439 | 16,781 | 1.3 |
| 1988 | 297 | 4,320 | 6.4 | 1,084 | 20,113 | 5.1 |
| 1989 | 243 | 3,671 | 6.2 | 1,551 | 20,882 | 6.9 |
| 1990 | 224 | 3,563 | 5.9 | 2,282 | 9,203 | 19.9 |

*Source*: compiled from data in FC Annual Reports.

initially operated alongside the Forestry Grant Scheme, which was geared more directly at commercial afforestation. The two schemes were merged (and modified) under the WGS.

Separate grant schemes and higher rates of grant for broadleaved planting are not the only manifestation of the growing opposition to the type of the new forests being created in Britain. The only native conifer grown on a significant scale is the Scots pine (*Pinus sylvestris*). Exotic species offer faster growth rates, especially in the moister western part of Britain, and some of them can also be used over a wide range of conditions. Sitka spruce (*Picea sitchensis*) in particular emerged as the most-used species (Table 2.3), and increasingly attracted criticism, especially from conservation interests. Another exotic species, Lodgepole pine (*Pinus contorta*) has also been extensively used on some of the poorer sites, and has been found to be susceptible to outbreaks of pests.

A feature of afforestation trends over the last few years has been the increased planting of broadleaved species (Table 2.4). A minimum content of 5 per cent broadleaved species is now expected in grant-aided afforestation projects, and many developments, especially on agricultural land in the lowlands, are exclusively broadleaved. Broadleaved species are, of course, usually characterised by slower growth rates. The trend towards the use of broadleaves reflects a shift away from the primacy of timber production in British afforestation in recent years, and reflects a significant underlying change in priorities of the objectives of afforestation.

The introduction of annual management grants, payable from 1992, also reflects the same policy shift. Payments will be offered for forms of management that will increase the environmental value of the woodland, or improve the provision of public access and recreational facilities.

### The spatial pattern of afforestation

From the time of the Acland Committee it was assumed that afforestation would be restricted to non-arable land, and indeed after World War II it was restricted to the poorest grades of hill grazings. This policy, as well as having obvious implications in relation to the choice of tree species, also had strong spatial consequences. In effect, afforestation was restricted to the uplands of the north and west, and as the more favourable and better located parts of the uplands were planted it shifted progressively northwards. South-west Scotland emerged as a preferred area, offering rapid growth rates and extensive areas of plantable land. As it and other optimal areas were progressively planted, the search for plantable land moved north, reaching the north coast of Scotland by the 1980s.

Issues of landscape conservation further influenced the spatial pattern of afforestation. National parks occupy around one-third of the upland area of England and Wales, and opposition to afforestation in these areas strengthened during the 1960s and 1970s. This is one of the reasons why an increasing proportion of British afforestation was concentrated in Scotland, where there are no national parks (Table 2.2). The process culminated in the announcement in 1988 that, in effect, coniferous afforestation on a commercial scale would no longer take place in the English uplands.

As has been indicated, from World War II until the mid-1980s, afforestation tended to move 'up the hill', on to progressively poorer land in the north and west of Britain. Over the last few years, however, there are signs of a complete reversal. In an age of agricultural surpluses, it no longer makes sense to maintain a rigorous policy of protecting farmland against afforestation. The reversal of earlier policy is reflected in the very substantial 'Better Land' supplements now payable on planting grants. Whereas in the past all but the poorest areas of farmland were protected against afforestation, the afforestation of (non-prime) arable land is now encouraged. This shift in policy coincided in time with the removal of tax benefits and a fall in planting rates, and it is therefore too early to assess its long-term significance. It may be expected, however, that a shift 'down the hill' will reduce pressures of afforestation in the uplands and mountains of the north and west, and that it may lead to the emergence of a different type of afforestation, involving better land and different species, in the lowlands.

## Social aspects of afforestation

From the time of the Acland Report, afforestation was perceived as offering social benefits in the form of the provision of employment and as a means of helping to stem rural depopulation. Its record in this respect has, however, been mixed. Policies of employment provision have been compromised by quests for economic efficiency, including changing intensities of management. The introduction of mechanisation meant that labour intensities per unit area declined drastically (as they also did in many forms of agriculture). Many of the new forestry villages established in remote rural areas have not survived in their intended form (Wonders 1990): in many instances their houses have been sold as second homes to urban-based purchasers.

Little evidence has been found to support the view that afforestation has significantly helped to halt rural depopulation, and indeed some claims have been made that it may accelerate depopulation in areas previously under small-scale farming. In a study of Snowdonia in north Wales, no indication that afforestation actually caused depopulation was found by Johnson and Price (1987), but they concluded that the break in job provision between planting and harvesting meant that forestry was unable to provide a stable economic base on which to maintain population.

While it is true that forestry-based employment in Britain increased by 4 per cent between 1986 and 1989, most of the expansion was in processing industry (often urban-based). Employment in establishment and maintenance fell by 18 per cent during that period, while the annual planting area increased by 20 per cent (Thompson 1990). In Strathdon in north-east Scotland, the forest area increased by 93 per cent over a 30-year period, but local employment in forestry and (game) keepering decreased by the same percentage (Evans 1987).

Most of the afforestation-related employment in recent years has been in the form of mobile squads rather than local residents. These squads, contracted by the forestry management companies, sometimes cover considerable distances. Long-established estate-based forests often have employment intensities that are higher than those of alternative land uses such as hill sheep farming. It remains to be seen whether more recently established forests will yield similar comparisons.

Since incentives for afforestation were in the form of tax relief to high-rate tax payers with large taxable incomes, farmers and small land owners were unlikely to be involved. No tradition of farm–forest integration exists in Britain, at least partly because of the early and extensive deforestation. In addition, the landlord–tenant system that dominated agriculture until the twentieth century militated against the direct involvement of farmers in forestry. The long interval between initial investment in afforestation and return on harvesting had a similar effect.

By the second half of the 1980s, however, there were signs that the traditional gulf between farmers and forestry was beginning to narrow. The removal of tax benefits and the compensating increase in the levels of planting grants favoured farmers and small land owners (as opposed to the previously dominant private investors). The relaxation of policies of protection of all but the poorest agricultural land meant that integration of farming and forestry became physically easier. And the introduction of a Farm Woodland Scheme (FWS), with annual payments for 20 or more years depending on species planted, in addition to planting grants, helped to make afforestation more attractive to farmers. An annual target totalling 12,000 ha per year was set for the first three years of the Scheme commencing in 1988, but achievement fell far short. By 1991 less than 12,000 ha had been planted.

Whilst the results of the FWS may have been disappointing in terms of the area planted, there is no doubt that the introduction of the FWS marked the beginning of a new era. Woodlands created under the FWS are typically small, broadleaved in composition, and intended to provide benefits of amenity, shelter or sport rather than commercial timber. Their contribution to overall forest expansion may be limited, but they are likely to lead to greater diversity in lowland agricultural landscapes and may in the long term lead towards closer integration of farming and forestry.

## Policy review

Radical reviews of policy, of the kind found in the Acland Report and in *Post-war forest policy*, have been conspicuously absent in recent decades. Policy statements – as represented, for example, by the Conservative government's statement in 1980 – have been common enough, but they have almost invariably been partial rather than comprehensive. Whilst enormous changes have taken place in afforestation policy even during the last decade, they have been incremental rather than radical, and lack clear signs of integration or co-ordination.

It is not surprising, therefore, that a fundamental reappraisal of policy has been urged by various groups, including Parliamentary committees (House of Lords 1980; House of Commons 1990), government-related agencies such as the Countryside Commission (1987, 1990) and interest groups such as the Council for the Preservation of Rural England (Stewart 1987) and Scottish Wildlife and Countryside Link (1992). In an attempt to respond, a policy statement was issued in 1991, but it was very sparse on detail. Policy objectives were defined as follows: 'First, the sustainable management of our existing woods and forests, and second, the steady expansion of tree cover in Britain to increase landscape beauty, wildlife habitats, public recreation, carbon storage, and of course to provide a source of timber' (Hansard Written Answers, 6 November 1991). Whilst the 'greening' of forestry policy is clear, many questions remain to be answered about the priorities and relative importance attached to the individual components of the second objective. As the Forestry Commission has indicated, 'Recent Government policy statements have placed increasing emphasis on the multi-purpose management of woodland, with all that this can provide in economic, social and environmental terms' (FC 71st Annual Report 1991, p. 10). But unless 'multi-purpose' policy is elaborated or disaggregated in terms of priorities, targets and locations, there is a danger that uncertainty will persist.

The changes that have occurred in recent years have perhaps further emphasised the disjointed and *ad hoc* nature of policy: for example, the planting target was increased from 30,000 to 33,000 ha in 1987 (in the context of the perceived need to reduce the agricultural area at a time of agricultural surpluses), whilst the major incentive for private planting (tax benefits) was removed in 1988.

The late 1980s and early 1990s will in the future probably be seen as marking a major turning point in the history of afforestation in Britain. With a doubling of the forest area since the beginning of the century, it is undeniable that an impressive form of progress has been achieved. Equally, however, it is apparent that the forms of afforestation carried out in previous decades are no longer acceptable. The primacy of timber production, reliance on one or two exotic coniferous species, and the modes and patterns of afforestation have all come under attack. While other objectives such as the provision of recreational opportunities and environmental benefits had been grafted on to the primary objective of timber production, that primary objective was itself weakening.

Criticism of afforestation policies on economic grounds was frequently voiced during the 1970s and 1980s (e.g. MAFF 1972; NAO 1986): rates of return on most new planting were often so low as to fail to meet the standard criteria for government investment. Much depends, of course, on the standpoint from which evaluation is being made, and on assumptions about non-timber benefits as well as about timber prices. Forestry management companies during their heyday in the 1980s indicated real rates of return of 4–6 per cent on private investment. Recent analyses have suggested maxima of around 6 per cent for conifers in the uplands, if other benefits are included and favourable asumptions made about land values, whilst native broadleaves in the lowland are likely to yield very much less (Pearce 1991).

Private investment prospects will increasingly be viewed within the wider international context, with overseas locations perhaps offering shorter rotations, faster growth rates and possibly fewer constraints of environmental planning. Whilst it is the case that Britain is still only around 12 per cent self-sufficient in wood-based products, and these products are a major item on the import bill, it does not necessarily follow that further afforestation within Britain is justified on economic grounds.

The changing outlook is perhaps epitomised by talk of creating new forests in lowland areas around some of the major cities and in the English Midlands and central Scotland (e.g. Countryside Commission 1987). These forests would be quite different from the extensive coniferous plantations established in parts of the uplands. They would, for instance, contain high proportions of broadleaved trees and open spaces, and would be established at least as much for their environmental and amenity benefits as for timber production. The increasing emphasis on such forests was reflected in the introduction of a large planting-grant supplement in 1991, aimed at creating new woodlands near large centres of population with few alternative opportunities for woodland recreation. This marks a further stage in the move away from 'timber primacy', and may perhaps encourage a further shift 'down the hill' and towards the south and east.

This is not to say that commercial coniferous afforestation will not continue, perhaps at a slower rate than in the past and subject to stricter environmental and planning controls. Such afforestation, however, is less likely to be driven by tax incentives than that of recent decades, and to be less strongly concentrated in the hands of 'absentee' private investors and of forestry management companies operating on their behalf.

Major changes have occurred in specific aspects of afforestation policy in Britain in recent years, but they have tended to be *ad hoc*, fragmented and incremental. Perhaps the time has come for a wide-ranging review of policy objectives in terms of the amount, type and spatial pattern of afforestation considered desirable in Britain at the end of the twentieth century. During this century, the forest area has expanded greatly as a result of direct state activity and of indirect state incentives. The underlying objective was timber production, and most of the afforestation was coniferous in character and was mainly located in the remoter north and west of Britain. The outlook for the foreseeable future is for forest expansion to continue, but probably at a slower rate than in the past. Most of the planting will probably be in the private sector, stimulated and directed by planting grants. More of it will probably be broadleaved in character, and increasing proportions are likely to be in the lowlands and around the big cities. Timber production will be less dominant as an objective of policy, and environmental benefits and recreational opportunities will have correspondingly higher priorities.

## Note

Statistical data are from the Annual Reports of the Forestry Commissioners, unless otherwise indicated. Except where otherwise indicated, policy statements are from House of Commons Parliamentary Debates (Hansard).

## References

Acland Committee (1917–18) *Final report of the Forestry Sub-committee of the Reconstruction Committee*, Cd 8881. HMSO, London.

Anderson, M.L. (1967) *A history of Scottish forestry* (2 vols.). Nelson, Edinburgh.

Brotherton, I. and Devall, N. (1988) Forestry conflicts in the national parks, *Journal of Environmental Management* 26: 229–38.

Caird, J.B. (1980) 'The reshaped agricultural landscape', in M.L. Parry and T.R. Slater (eds.) *The making of the Scottish landscape*. Croom Helm, London, pp 203–22.

CAS (Centre for Agricultural Strategy) (1980) *Strategy for the UK forest industry*. CAS, Reading.

Countryside Commission (1987) *Forestry in the countryside*. Countryside Commission, Cheltenham.

Countryside Commission (1990) Evidence, in 'Land use and forestry', Second Report of the House of Commons Agriculture Committee, *House of Commons Paper 16*. HMSO, London.

Departmental Committee (1902) *Report of the Departmental Committee appointed by the Board of Agriculture to enquire into and report upon British forestry*, Cd 1319. HMSO, London.

Evans, A. (1987) The growth of forestry and its effects upon rural communities in north-east Scotland: the case of Strathdon, *Scottish Forestry* 41: 310–13.

Forestry Commission (1977) *The wood production outlook in Britain*. FC, Edinburgh.

Forestry Commissioners (1943) *Post-war forest policy*, Cmd 6447. HMSO, London.

Hetherington, M. (1988) Afforestation consultations in Northern Scotland: a case study of the voluntary system in operation, *Scottish Forestry* 42: 185–91.

House of Commons Agriculture Committee (1990) *Land use and forestry*, House of Commons Paper 16. HMSO, London.

House of Lords Select Committee (1980) *The scientific aspects of forestry*, House of Lords Paper 381. HMSO, London.

James, N.D.G. (1981) *A history of English forestry*. Blackwell, Oxford.

Johnson, J. and Price, C. (1987) Afforestation, employment and depopulation in the Snowdonia National Park, *Journal of Rural Studies* 3: 195–205.

MAFF (Ministry of Agriculture, Fisheries and Food) (1972) *Forestry policy*. HMSO, London.

Mather, A.S. (1987) The structure of forest ownership in Scotland, *Journal of Rural Studies* 3: 175–182.

Mather, A.S. (1991a) Planning and afforestation in Scotland: the nature and role of afforestation consultations, *Environment and Planning A* 23: 1349–59.

Mather, A.S. (1991b) The changing role of planning in rural land use: the example of afforestation in Scotland, *Journal of Rural Studies* 7: 299–309.

Mather, A.S. and Murray, N. (1986) Disposal of Forestry Commission land in Scotland,

*Area* 18: 109–16.

Mather, A.S. and Murray, N. (1988) The dynamics of rural land-use change: the case of private-sector afforestation in Scotland, *Land Use Policy* 5: 103–20.

NAO (National Audit Office) (1986) *Report by the Comptroller and Auditor-general: review of Forestry Commission's objectives and achievements*, House of Commons Paper 75. HMSO, London.

NCC (Nature Conservancy Council) (1986) *Nature conservation and afforestation in Britain*. NCC, Peterborough.

Pearce, D. (1991) *Forestry expansion: a study of technical, economic and ecological factors: Paper 14 – Assessing the return to the economy and to society from investments in forestry*. Forestry Commission, Edinburgh.

Royal Commission (1909) *Second report (on afforestation) of the Royal Commission appointed to inquire into and to report on certain questions affecting coast erosion, the reclamation of tidal lands and afforestation in the United Kingdom*, Cd 4460. HMSO, London.

Scottish Wildlife and Countryside Link (1992) *A forest for Scotland*. SWCL, Perth.

Select Committee on Forestry (1887) *Report*. HMSO, London.

Stewart, P.J. (1987) *Growing against the grain: United Kingdom forestry policy*. Council for the Protection of Rural England, London.

TGUK (Timber Growers UK) (1985) *Forest and woodland code. TGUK*, London.

Thompson, J. (1990) Forest employment survey 1988–89, *Forestry Commission Occasional Paper 27*. Forestry Commission, Edinburgh.

Watkins, C. (1986) Recent changes in government policy towards broadleaved woodland, *Area* 18: 117–22.

Wonders, W.C. (1990) Forestry villages in the Scottish Highlands, *Scottish Geographical Magazine* 106: 156–66.

Zuckerman Committee (1957) *Report of the Natural Resources (Technical) Committee: forestry agriculture and marginal land*. HMSO, London.

# 3 Afforestation in the Republic of Ireland

*Desmond A. Gillmor*

## Introduction

Afforestation in the Republic of Ireland has a unique role and distinctive characteristics, although the country is the least forested in Europe apart from Iceland. Forests occupy only 6 per cent of the land, but this represents six times the area of forest at the time of political independence in 1922. The area of natural woodland is negligible, so that the present forest cover owes its origin essentially to afforestation. This afforestation constitutes the most striking land-use change that has been occurring in rural districts, and it is having a major impact on the Irish countryside. The rate of afforestation per capita is greater than in any other country. There has been an unusually high level of state participation in this afforestation, so that 85 per cent of the country's forest is publicly owned, twice the proportion in the European Community (EC) as a whole, but this situation is currently changing.

Unlike in Iceland, adverse environmental conditions cannot be invoked as a reason for the paucity of forests in Ireland. The temperate and moist climate combines with suitable soils to promote rapid tree growth. While unqualified international comparisons are dangerous, Irish yields are the highest in Europe and are treble the EC average. The mean annual growth increment of coniferous plantations is 15 cubic metres per hectare, with more than half the country being in the very high yield classes of 18+. Irish forests are relatively free from the effects of acid rain, pests and diseases and the incidence of fire damage is low. Detracting somewhat from the environmental advantage of the highest productivity sites on wet mineral soils are liability to windthrow because of shallow rooting, and some impairment of timber quality by rapid growth.

Because of the suitability of the environment for their growth, forests comprised the natural vegetation over much of Ireland. However, they gradually gave way to farmland through the effects of grazing, cutting and burning, as human settlement and agriculture expanded. The process of forest clearance resulted also from the spread of peat bogs and from commercial exploitation for lumber, charcoal and tanbark. The protective management of English forests was not extended to Ireland, and depletion of the forest resource was almost complete by the eighteenth century (Fitzpatrick 1966; McCracken 1971; O'Carroll 1984; Neeson 1991).

**State afforestation**

Efforts to replenish the Irish forest resource have been largely through afforestation by the state. State forestry had been proposed in an Afforestation Bill introduced to the British Parliament by an Irish member in 1884 but this bill was unsuccessful. Reports on Irish forestry prospects were commissioned, however, and these led to the first attempt at silviculture by the state in the British Isles. It was known as the Knockboy Experiment of 1890–98 and it involved planting on a 400 ha site in County Galway offered by the local parish priest. It was a dismal failure because of the nature of the rocky site on the wind-swept west coast and the use of inappropriate species. While drawing attention to the idea of state afforestation, the failure reinforced the dismissive attitude towards forestry on the part of some in authority and contributed to a cautiousness in forestry which was to last for decades.

Despite Knockboy, effective state afforestation in Ireland began ahead of that in Britain, with the purchase of the Avondale Estate in County Wicklow in 1904 and establishment of a forestry school there two years later. This reflected the continued enthusiasm of many individuals and the influence of the Irish Forestry Society which had been founded in 1900. The optimistic perception of Ireland's forestry potential was expressed officially and most influentially by the Departmental Committee on Irish Forestry (Department of Agriculture 1908). It proposed that the state should plant 80,000 ha of waste land as part of an ambitious target of 400,000 ha of forests in Ireland. While this report had a major impact on later thinking, achievements in the short term were more modest than projected, because of financial limitations and the intervention of wartime conditions. About 1300 ha had been planted and a further 9000 ha of land acquired by the state at the time of establishment of the independent Irish Free State in 1922.

Belief in the desirability of state afforestation had been associated with the rise of Irish nationalism, but this was slow to translate into effective forestry policy and commitment in the decades after independence. This seems to have been at least in part because politicians, faced with the realities of scarce finance and electoral accountability, favoured investments which could be seen to yield more immediate results. Much depended upon the interest and commitment of individual politicians and administrators, as when, following a change of government in 1932, the enthusiastic Joseph Connolly became the responsible minister in 1933. Annual planting had increased from 400 ha in 1922 to 1400 ha in 1928 but stabilised thereafter until the annual planting target was raised from 1400 ha to 2400 ha in 1933 (Figure 3.1). The rate of planting doubled within the next two years and reached 3100 ha by 1938, facilitated by the greater ease of acquiring land because of the agricultural depression of the 1930s. This gain was lost during World War II, when activity was hindered by difficulties in obtaining seed and fencing material, and attention within forestry shifted towards felling in response to the urgent demand for timber. The total area of state forest in 1948 was 53,000 ha and the ultimate national target for state and private forest had been set at 283,000 ha.

The year 1948 marked a watershed in Irish state forestry. This was because of the announcement of a proposal to increase progressively annual planting by the state to 10,000 ha, towards a target of 400,000 ha of forest additional to that

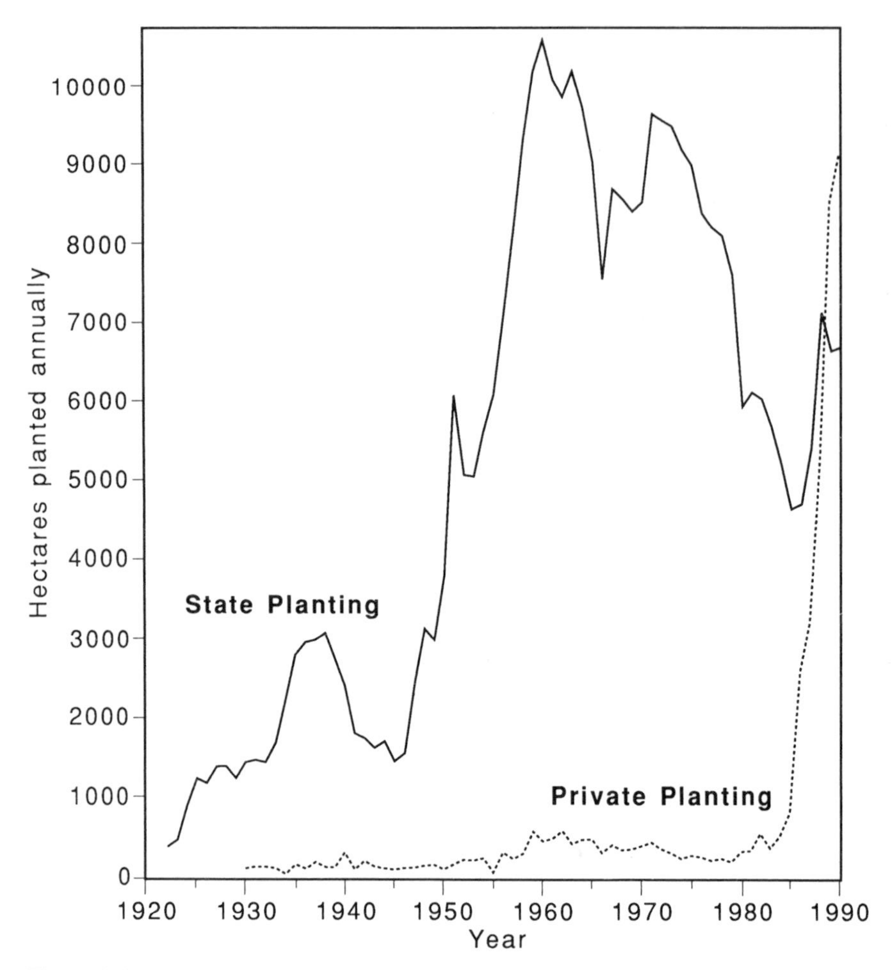


**Figure 3.1** Afforestation in the Republic of Ireland, 1922–1990.

existing at the time. Individual foresters had been calling for some time for major expansion of afforestation and there was increased media and public interest in forestry. This was stimulated by the serious wartime depletion of timber resources. It is unlikely, however, that such an ambitious programme would have emerged without individual influence in the context of the particular political circumstances of the time. Seán McBride had a passionate interest in afforestation and he made its expansion a condition of his participation in a coalition government in 1948, having earlier broken with the previous government party and established a new one because of its failure to adopt his forestry policy. Although, as Minister for Foreign Affairs, he did not have responsibility for forestry in the new government, he succeeded through skill and perseverence in having his afforestation policy implemented against much opposition, including that of the Department of Finance. The change from a gradualist policy to one of vigorous expansion led to an increase in annual planting from 3000 ha in 1949 to attainment of the 10,000 ha target ten years later. This was greatly facilitated by the possibility of extending afforestation on to more difficult land after new machinery enabling deep ploughing became available in the early 1950s.

Annual state planting averaged over 9600 ha in the period 1959–75, fluctuations generally reflecting the land-acquisition situation. With agricultural development and EC membership, the price of land increased and farmers became more reluctant to sell it for afforestation, and later there was national recession and curtailment of government expenditure on forestry because of budgetary difficulties. The reserve of state land suitable for afforestation was depleted and the rate of planting declined to 6300 ha in 1985. The annual target had been reduced to an interim one of 7500 ha, which was now to include reforestation of land previously planted by the state and private forestry. Statistics for such reforestation are not included in Figure 3.1 and were published separately first for 1961, when 200 ha were planted, but there had been little previously because the small area of early state forest was just coming to maturity. Reforestation increased gradually thereafter but rose more steeply from 700 ha in 1980 to 1700 ha by 1985. Thus reforestation was assuming an increasing role within a diminishing state planting programme, so that the 4600 ha of afforestation achieved in 1985 was the lowest since 1950.

Subsequently there has been an upturn in the prospects for state forestry. This has been an outcome of forestry being accorded much greater priority in interrelated national and EC policies. Considerable importance was attached to the potential for forestry in national development programmes in 1987, 1989 and 1991. This was related to the increasing recognition and encouragement of forestry as an alternative land use to agriculture within the EC because of surplus farm production and timber-supply deficits. The reform of the structural funds associated with the Single European Act provided greatly increased aid to Ireland and this afforded the opportunity for substantial EC support of Irish forestry. A five-year Forestry Operational Programme 1989–1993 (Government of Ireland 1991) was approved by the EC and is one-third financed by it. It provides for supporting an increase in state afforestation from 6700 ha in 1989 to 8100 ha in 1993, with 65 per cent grant aid to the semi-state forestry company for approved costs up to certain limits. The area afforested by the state in 1990 was 6670 ha, accounting for 64 per cent of the state planting of 10,352 ha in that year, the remainder being reforestation. The total area of state forests was then 360,000 ha, occupying 5.2 per cent of the land in the country and making the state forestry service by far its largest land owner.

State forestry was until recently within the civil service as part of a government department and controlled by the Forestry Act 1946 (McEvoy 1979). There had often been suggestions that it should be transferred to the semi-state sector, so that it might have a more commercial orientation and less bureaucratic control. In 1984 the government initiated a study of the then Forest and Wildlife Service. The resultant report of the Review Group on Forestry (Department of Fisheries and Forestry 1985) recommended restructuring of the service within the civil service to give a commercial commission which would be called National Forest Enterprise. Three years later, however, a semi-state Irish Forestry Board, titled Cóillte Teóranta, was established by a new government under the Forestry Act of 1988. Cóillte was given responsibility for the commercial operation, development and diversification of the state's forestry activities on a profitable basis.

## Private afforestation

With the natural forest of Ireland almost gone, the first attempts to reverse the trend of resource depletion came through a phase of estate forestry in the eighteenth and nineteenth centuries. Landlords planted trees on their estates, at first mainly for aesthetic effect and later with commercial motives. The peak of this private afforestation was passed in the 1870s, as the subsequent abolition of the landlord system through land reform was associated with large-scale woodland clearance and deterioration. This private estate forest accounted for most of the country's forest at the time of independence but it has declined since by a half, to less than 50,000 ha, mainly through acquisition by the state. The remainder is of mature age structure and generally deteriorated and understocked condition. Yet because it is mainly deciduous, estate woodland makes an important contribution to the aesthetic quality of the landscape in many places. Although the seminal report of the Departmental Committee on Irish Forestry (Department of Agriculture 1908) had envisaged that three-quarters of the afforestation programme which it proposed the state should organise would be through private forestry, the outcome has been very different. The principle of providing state grants towards private forestry was adopted in 1928 and data relating to this provide the only basis on which to trace private planting (Figure 3.1). The response to this incentive was slight, exceeding 100 ha nationally in only one year during the 1930s. There was steady but still very slow growth in the post-war period, some of the planting then being because of a requirement that felled forest should be replaced. It had been concluded in an FAO (1950) report that private forestry could not be expected to make a significant contribution to the national timber supply. A campaign to promote private afforestation initiated in 1958 under E. Childers as minister led to a significant but short-lived response, with annual planting falling from a peak of 530 ha in 1962. Promotion had included doubling of the state grant per hectare, provision of free technical advice and extensive publicity. The level of planting had fallen to 130 ha by 1979 and the total extent of grant-aided private planting from 1929 did not exceed 10,000 ha until 1981. This was equivalent to only 3.4 per cent of the forest which had been planted by the state at that time.

Any attempt to explain the very minor role which private planting played in afforestation in the Republic of Ireland since independence must involve consideration of varied influences. The small size of most farm holdings ensured that land was at a premium and this was an obvious barrier to owners devoting space to trees. If the possibility of afforestation were considered, the costs and labour incurred in establishment, the risks involved, and especially the long time-scale of any return on investment and of income expectation, would have been major disincentives. For those receiving social welfare payments, there was the fear that income from forestry might detract from their entitlement. Most land owners had no familiarity with or knowledge of forestry and adequate information concerning its potential was not readily available to them. There was no forest consciousness amongst people in general, perhaps attributable at least in part to a tendency to associate forestry only with the former landlord class and later with the state. This may have been reinforced by the feeling that to grow trees would be a waste of agricultural land and would be seen by others as admission of failure as a farmer. Many of the remaining small former landlord class who had an appreciation of forestry had lost much of their estates, were antagonised by the land reform, were in reduced economic circumstances and were unsympathetic to the new political situation, so were not prepared to undertake afforestation.

Explanation for the predominance of state over private afforestation must be sought also in state policy. Although private planting was encouraged and incentives were offered, no clear policy was formulated for it and there was little evidence of a strong commitment, apart from that of a few individual foresters and politicians. This may have been related in part to a nationalist viewpoint in seeing the development of the country's forest resources as being the function of the new state itself, reinforced by the traditional association of private forestry with what were perceived as alien landlords (Neeson 1991). Active commercial private afforestation would have been considered as a potential competitor for land with the state forest service. For this reason, private afforestation was seen for long only as farm forestry, and grant aid would not be given to companies which would be purchasing land for forestry. With regard to farmland, agriculture was considered the desirable use where feasible, and forestry administration was within government departments where agricultural interests predominated from independence until 1977. The prime aim of private forestry promotion seems to have been limited to encouragement of the planting of small areas unsuitable for agriculture on farms.

The environment for private afforestation in the Republic of Ireland altered fundamentally in the 1980s. Although this was related in part to affairs within the country, the main influence on change has been the EC. With the decline in state planting, the government had begun to look more towards private afforestation in order to meet national targets and to do this more cheaply. This emphasis increased with the general trend towards lessening direct state participation in the economy and promoting private-sector involvement. The favourable investment potential in Irish forestry had been emphasised in various studies and reports, such as those by the National Economic and Social Council (1979) and the Forest and Wildlife Service (1980). These opportunities began to attract the attention of financial institutions, especially those, such as pension funds, seeking longer-term secure investment. A major incentive was provided under an EC Western Package Programme for ten years from 1981, which included substantial grant aid towards planting in twelve western counties. Yet farmers in particular were slow at first to avail themselves of this opportunity and only one-tenth of the target expenditure was taken up in the early years. The EC commitment to forestry developed as increasing agricultural surpluses prompted efforts to divert farmland to alternative uses, with forestry in contrast having particular attraction because of the current and projected timber deficit. The apparently poor prospects for the main farm enterprises have been a major incentive for farmers to re-evaluate their attitudes towards forestry and to consider it as a viable alternative. Its attraction for some had increased with assurances that forestry income would not detract from social-welfare entitlements. From 1986, annual livestock headage payments to farmers in disadvantaged areas who converted to forestry were continued for fifteen years. This measure was later succeeded by an annual forest premium payable to all farmers throughout the country, thus having the very important effect of alleviating the short-term income problem. In 1988 aid under the Western Package was extended to the disadvantaged areas in all parts of the state and the EC contribution to grants was raised from one-half to two-thirds. It has been succeeded throughout the state by the Forestry Operational Programme 1989–1993, which is a further major boost to private afforestation in providing that it would increase to 14,200 ha by 1993, when private planting would account for 64 per cent of all afforestation. Grants are paid up to 85 per cent of approved costs for farmers and 70 per cent for others, within certain limits. Not only is the planting of agricultural land eligible for grant aid but a higher level applies to it. Investment is encouraged by forestry being exempt from income tax and corporations' profits tax, and it also receives favourable treatment in the assessment of capital acquisitions, though investment in forestry cannot be offset against tax liabilities elsewhere. The organisation of planting through the establishment of commercial forestry companies and co-operatives has greatly facilitated development.

Although the full impact of these changes had not yet been felt, the data for planting to 1990 indicate the transformation in private afforestation that had occurred (Figure 3.1). Planting rose only gradually during the first half of the 1980s, but it escalated from 760 ha in 1985 to 9200 ha by 1990. Private afforestation of 28,800 ha over the five years 1986–90 was more than double all private planting in the preceding 55 years. That done in 1990 amounted to 58 per cent of total afforestation in that year. Farmers accounted for 45 per cent of the private forest planting. Afforestation in the Irish Republic had embarked upon a major new phase of development.

## Objectives of afforestation policy

The policy of afforestation in the Republic of Ireland resulted from a blend of objectives which varied over time but which were not clearly enunciated. For a long time the most specifically stated and apparently prime objective was national self-sufficiency in timber supply. This had a limited defence-strategy element but a more persistent reason has been import substitution to benefit the national balance of payments, reinforced perhaps by prestige considerations of the developing state having its own forests. Domestic sources will have effectively saturated the Irish timber market at 75 per cent by the mid-1990s. Attention to the foreign-exchange earning potential of forestry followed from a reorientation of national economic policy towards export promotion from the 1960s, with one-third of production being exported in 1990, mainly to the United Kingdom.

There has always been the idea that forestry should be an economically viable form of rural land use and a rational national investment. It was only with the adoption of commercial management techniques in the 1950s, however, that economic efficiency became an objective of policy and the emphasis on profitability has increased in recent times. Because it is still based largely on afforestation, the capital costs in the Irish forestry industry are high and the state forestry service recorded an operating profit for the first time in 1990. Estimates of profitability are fraught with difficulty, including assumptions about price and discount rates in the long term, but projections since the 1960s suggest real annual returns in the range of 2–6 per cent, with the current aim of financial institutions investing in afforestation being at least 5 per cent (O'Hegarty *c.*1989).

There has always been recognition of the employment benefit of afforestation but the relative emphases on commercial and social objectives have varied, and there has been tension between the two. Its role has been seen as that of affording remunerative employment in rural areas, widening the range of jobs and increasing skills there, and lessening depopulation of the countryside. Employment needs have been all the greater in the more deprived western part of the state, so emphasis on afforestation there has a regional-development dimension. Direct state forestry employment increased in line with planting rates from less than 500 in 1930 to a peak of 5000 in the 1950s but, with improving labour productivity, has since fallen to 2000, representing one job per 180 ha. All employment projections have over-estimated the contribution of forestry but staff costs account for half of the expenditure by Cóillte, so that it is still an effective way of providing employment. There is greater indirect employment in the harvesting, transport and processing of timber.

Attention has been given since the 1950s to the contribution which afforestation can make to manufacturing development through processing of the increasing forest output. Four pulpwood plants were established but all closed with the reduced demand and increased competition of the recessionary period 1978–83. The largest reopened later and a new medium-density fibre plant began production in 1984.

A prime objective permeating afforestation policy until the 1980s was that competition with agriculture, especially for land, should be minimised. Assessments of comparative profitability with agriculture indicate a substantial relative attractiveness of forestry on low-quality land, especially on wet mineral soils. Thus the policy of not using land which might be considered suitable for agriculture was not based on productivity criteria but was to protect the interests of farming and avoid conflict for social and political reasons. This was implemented through the financial control of adherence to low land-acquisition prices and the institutional regulator of consultation between the forest service and the Irish Land Commission, which was the body responsible for agrarian structure.

Recreational provision became an objective of state forest policy only from the late 1960s. This relates only to public access to established forests and no afforestation has been undertaken for recreational purposes. Provision grew rapidly and 400 sites are open to the public, ranging in amenity provision from twelve forest parks to places with only basic forest walks.

The clearest statement of forestry objectives has been that made in the Forestry Operational Programme 1989–93 (Government of Ireland 1991). This stated that the prime objective is to contribute to the generation of wealth in the Irish economy by utilising available and suitable land, and human and financial resources, to the best advantage in creating new and developing existing forests so as to: provide the raw material base for an expanded and improved forestry-based industrial sector; diversify the rural economy; stimulate rural development; provide employment; and promote the reform of agricultural structure. Although not included in the programme list, the making of a positive contribution to the environment was added to the objectives in the preface by the minister.

## Forest characteristics

There are considerable variations in the suitability of Irish land for agriculture, the main distinction in very general terms being between the better land of the lowlands and of the east and south, and the poorer land of the uplands and of the west and north. The type of land available for afforestation has a strong influence on the spatial distribution of forestry and it interacts also with the species of trees used. The type of land planted, the spatial pattern of activity and the species of tree used have changed over time, so that the present character of the national forest results from an amalgam of influences from different times in its evolution (Marchand 1975; Gillmor 1985).

Most of the estates were on the lowlands and they were more numerous in the east and south, so that the environmental conditions were suitable for the broadleaved trees which the landlords generally favoured in the estate phase of afforestation (McEvoy 1979). One-third of state planting was on such former estate woodland prior to World War II. Some of this replacement was with deciduous trees, which comprised 11 per cent of state planting in the 1920s and 1930s.

Most afforestation has differed markedly from that on the former estates in being predominantly on land of lower agricultural potential and being comprised almost exclusively of conifers. The policy of not using land suitable for agriculture and the associated low prices paid for land for afforestation largely account for the spatial distribution of state forests, which is mainly in upland areas and on peaty or wet mineral soils on the lowlands (Figure 3.2). These environmental conditions are more readily tolerated by conifers, which are favoured also because of the large demand for softwood for industrial usage and because they mature within forty years.

Afforestation prior to the 1950s was mainly on mineral soils on the intermediate and lower slopes of the less exposed mountains which were considered safe for planting, principally in the upland districts of the east, south and midlands. Between the moorland of these ranges and the improved agricultural land, there was an extensive hill-pasture zone providing relatively favourable conditions for planting, which was further facilitated by the partial retreat of agriculture from the zone. Reliance was on European conifers, chiefly Norway spruce, Scots pine and larches. With unmechanised planting of Lodgepole pine on peat from the 1930s and a political wish at that time to promote planting in the west, foresters ventured on a small scale into more western areas, but the general approach was one of caution.

Several influences combined to bring about a major change in the emphasis in afforestation from about 1950, involving large-scale mechanised planting of peat and a rapid acceleration in the westward shift in afforestation. The huge increase in the annual planting target announced in 1948 required that there had to be a corresponding growth in land acquisition, at a time when competition for land with farming was beginning to increase, especially in the east. Following the

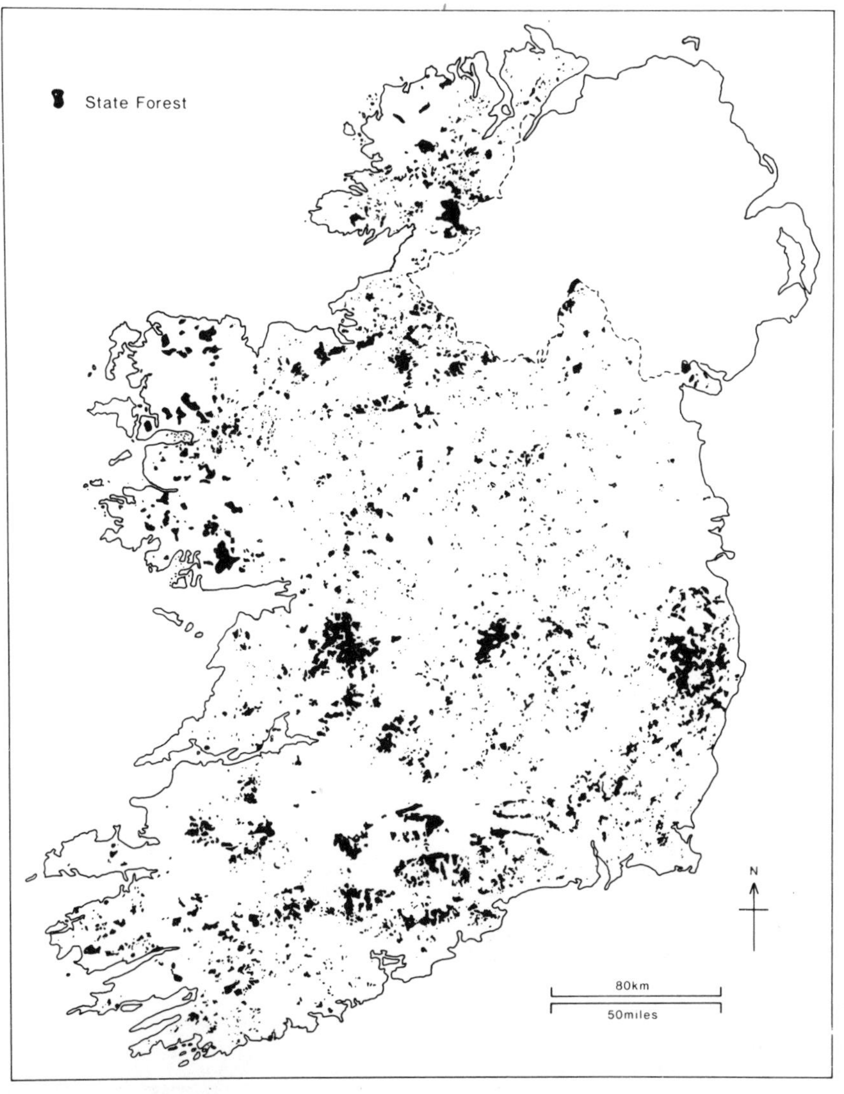


**Figure 3.2** The distribution of state forests.

success of British research on planting on deep peat using tractor ploughing and phosphatic fertilising, heavy machinery was purchased in 1951. Mechanisation permitted the immediate extension of large-scale afforestation on to the higher-rainfall blanket bogland and peaty podzols, involving a move upslope and westwards. This was facilitated also by a widening of the range of land that could be forested through substitution of North American conifers for the European varieties, research over preceding decades having shown them to be better adapted to Irish conditions, especially on the poorer, damper and more exposed sites. Sitka spruce became the leading species, followed by the use of Lodgepole pine on the

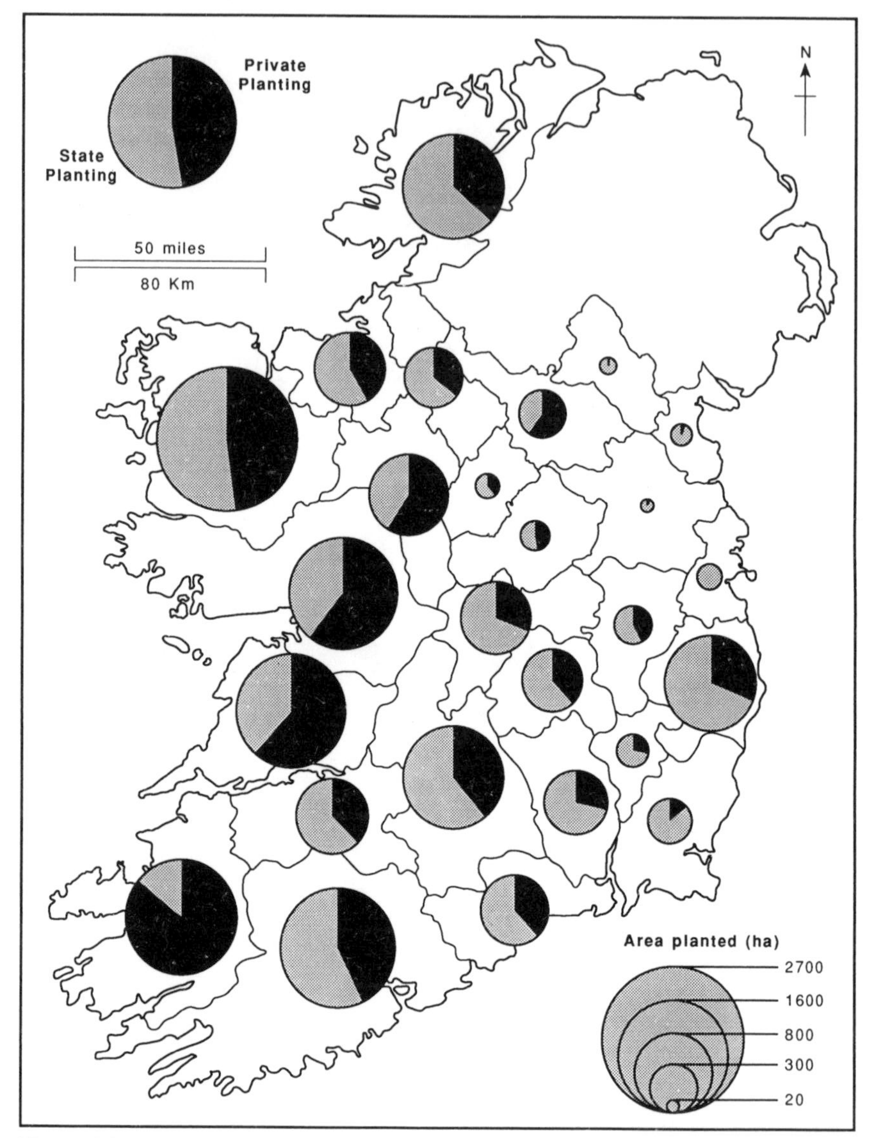


**Figure 3.3** Area of forest planted during the year 1990, by county.

poorest land and most exposed sites. The westward shift in the focus of afforestation was reinforced by the growth of interest in regional development in the 1960s and the perception of forestry as contributing significantly to social objectives of state policy. Two-thirds of the combined state and private planting in 1990 was in the west (Figure 3.3). Because of the later development of afforestation there, western forests have even younger age structures and less species diversity than the state forest as a whole.

Other changes began or accelerated in the 1980s, in association mainly with the interrelated factors of the growth in private afforestation, the greater commercial orientation in the state forest service and the changing relationship between forestry and farming. These have combined to initiate a reversal of the trend towards movement on to poorer land, as afforestation seeks more productive sites, including land which had been used for marginal agriculture. A major focus of attention is the 1.0 million ha of wet mineral soils, occupying 15 per cent of the land area, and lying mainly in the west and north, which are marginal for agriculture but are mostly of yield classes over 18 for Sitka spruce. Research from the early 1970s had demonstrated the potential of this land for forestry, but it was largely under a low-grade agriculture and incursion of afforestation was resisted. A different type of land which is receiving attention by Cóillte is the cutover peat bogs which are becoming available for afforestation as their energy resources are depleted. Raised bogs cover 400,000 ha, mainly in the midlands, and have yield class 20–24 for Sitka spruce. Some planting is being done on marginal soils on the lower hill slopes of the east and south and in pockets on the lowlands. The more exposed parts of the west are no longer being planted by Cóillte but even in the west some lowland blanket peats have yield range 14–20. The more commercial emphasis, however, now being reinforced by considerations of environmental conservation, is lessening the focus on the blanket peats of the west and higher altitudes elsewhere. This and some disappointment with results have contributed to a relative reduction in planting of the hardy Lodgepole pine relative to the dominance of the quick-growing Sitka spruce. The percentages of Cóillte planting in 1990 were Sitka spruce 76, Lodgepole pine 12, other conifers 10 and broadleaves 2. One-third of 'other conifers' was noble fir, planted as a result of the identification of potential export markets for Christmas trees. Sales of timber by Cóillte in 1990 were: pulpwood 37 per cent, large sawlog 36 per cent, small sawlog 26 per cent.

## Some issues in forestry development

While rapid afforestation is occurring in the Irish Republic within a favourable physical, policy and economic environment, there have been and are problems associated with its development. These issues, additional to some considered already, include those of attitudes, land acquisition and funding, commercial orientation and outlets, and amenity and species considerations.

Attitudes inimical to afforestation have hindered development at all stages and, while a major transformation has recently occurred, vestiges remain. Attitudes still range from local antagonism to seeing forestry as the panacea of rural areas. Political support was dampened in policy implementation by the long-term financial commitments associated with lengthy forestry rotations not complying with short-term electoral expediency. A major influence has been the traditional antipathy of the farming community based largely on competition for land, which has been such an important element in Irish rural society. While forestry is now generally viewed favourably in the contexts of national and regional development, there may be opposition to further afforestation in some of the rural areas affected. Its encroachment is resented by some farmers who fear diminution in sheep grazing and especially their prospects of obtaining land for farm enlargement. The latter, combined with opposition to the intrusion of commercial forestry interests, was the basis of conflict in County Leitrim in the 1980s which led to damage being caused to forestry machinery and plants. This was seen as a people-versus-forestry issue, with afforestation being referred to as a cancerous growth because it was perceived as heralding the decline of the traditional farming way of life, and was accompanied by dwindling population and farmstead dereliction.

Limitation of political commitment and competition with agriculture have been major causes of the perennial associated problems of finance and land acquisition for afforestation (Farrell 1983). Financial curtailment must be seen also in the context of the limited resources available to the state, especially in the past and still to some extent at present. Compulsory land-acquisition powers were acquired by the state in 1928 but have never been used except to facilitate the procurement of commonage. Instead land was acquired as it became available if it could be purchased at costs not exceeding the low fixed ceiling prices. This, associated with the small size of farms and tracts of land on offer, led to a fragmented forest pattern with many uneconomically small patches, some with poor accessibility. Substantial consolidation of state forest has been effected in recent years, in part through designation of prime acquisition areas.

Attitudes to forestry have been involved also in a traditional problem of inadequate commercial orientation and the associated lack of a marketing and processing dimension to policy. State afforestation was for too long perceived as an end in itself rather than as a commercial enterprise dependent ultimately on adequate and remunerative outlets for its produce. The availability of outlets has affected confidence in afforestation, as was seen in the beneficial influence of the opening of processing plants in the 1950s and the depressant effect of their closure about 1980. There is a need for greater product upgrading and development and for a co-ordinated plan of growth of manufacturing capacity as output expands. It is projected that sales of timber from state forests will increase from 1.4 million $m^3$ in 1990 to 3.0 million $m^3$ in the year 2000 and 9.0 million $m^3$ by 2030. The establishment of Cóillte, with strong commercial orientation and marketing objectives, has been a major development in countering deficiencies in the public sector. Its attempts to promote the establishment of a large pulpmill have so far failed and some timber is being exported unprocessed and as far as Finland. Irish forestry is far short of its potential in downstream development.

Amenity considerations have an important bearing on public attitudes towards afforestation. The image of forestry benefited greatly from the provision for and growth in forest recreation. In a study of forest visitors in 1976, 91 per cent felt that state forestry added to the appearance of the landscape and only 3 per cent considered its overall effect to be aesthetic detraction (Bagnall *et al.* 1978). Concern about the negative environmental effects of coniferous afforestation developed late in the Republic of Ireland, being expressed on a significant level only in the late 1980s. The impacts of forestry on Irish landscapes and habitats are accentuated by the extent of monoculture of exotic dark conifers in plantations which tend to have geometric layouts and by its concentration in open upland areas with frequently high scenic quality. Forests may often add variety and beauty to the scene but the distinctive regional diversity of landscape is lessened, attractive views have been obliterated and some unique areas should not have been planted. Concern has related mainly to the visual impact of afforestation and to the destruction of wild bogland habitats of conservation value (Dunstan 1985; Hickey 1990; Mollan and Maloney 1991). Other threats include the destruction of sites of archaeological, historical and scientific importance and the possible contribution to the acidification and siltation of rivers and lakes with consequent impacts on fisheries.

Little attention was given to the environmental aspects of afforestation in the early stages of development but the situation began to improve from the 1960s. It is only very recently, however, that a substantially greater recognition has been given, prompted at least in part by EC requirements and public concern. Environmental responsibilities were included in the Forestry Act 1988 and the Forestry Operational Programme 1989–1993. Guidelines have been formulated for acceptable forestry procedures with regard to water, landscape and sites of historical or scientific value. Compliance with these is required of grant recipients and of Cóillte, which has adopted its own environmental programme and is shifting its attention from virgin peatlands. It has initiated a policy of species diversification, which is projected to reduce Sitka spruce to 72 per cent of planting and increase hardwoods to 10 per cent, but this is being done for reasons of disease and pest vulnerability and of market supply as well as for environmental reasons. Diversification becomes more feasible as the quality of land used improves. Planting of broadleaved species is being encouraged by higher grant levels and it is promoted through the activities of the Tree Council of Ireland and an active voluntary organisation titled Crann. Forestry remains exempt from planning regulation except for areas exceeding 200 ha, for which environmental impact assessments are required under EC regulations. Improved standards of forest design are most desirable.

## Conclusion

Twentieth-century afforestation in the Republic of Ireland has made a substantial contribution towards reversing the forest-resource depletion of previous centuries, though the new forest is of very different type. The achievement seems all the greater when viewed against the background of the tiny forest industry at the beginning of the century, and the consequent dearth of a forest tradition. It took time to develop technical expertise, and expansion was hindered also by the importance attached to agriculture in Irish society and by the scarcity of funding available for long-term investment. Political commitment fluctuated but was often muted and a vigorous commercially-oriented policy was slow to evolve. The priority attached to forestry as a land use has been greatly enhanced within the EC policy context and afforestation is now on a much firmer footing. Future trends will be linked intimately to this policy and to the prevailing climate in agriculture. Half of the land in the state is marginal for agriculture and over half of this is suitable for forestry. Ireland has a considerable comparative advantage in forestry and further major extension of afforestation seems likely.

One great deficiency which has characterised afforestation policy in the Irish Republic and which still exists is the lack of forward planning with regard to the location of development. The present spatial pattern has evolved in a piecemeal fashion and the escalating private involvement suggests that this is all the more likely to continue. As a minimum initial requirement, an indicative strategy is needed to show through a system of zoning where afforestation of the appropriate type should occur. This should not be done in isolation but in the context of forestry as a multi-functional long-term land use which can contribute to the sustainable development of rural society and one which should be developed in harmony with the environment and with other land uses. Thus a national integrated land-use plan is needed to provide the framework within which there could be organised expansion of afforestation.

## References


Bagnall, U.E., Gillmor, D.A. and Phipps, J.A. (1978) The recreational use of forest land, *Irish Forestry* 35: 19–34.

Department of Agriculture and Technical Instruction for Ireland (1908) *Report of the Departmental Committee on Irish forestry*. HMSO, Dublin.

Department of Fisheries and Forestry (1985) *Review Group on Forestry.* Stationery Office, Dublin.

Dunstan, G. (1985) 'Forests in the landscape', in F.H.A. Aalen (ed.) *The future of the Irish rural landscape.* Department of Geography, Trinity College Dublin, pp 93–153.

FAO (1951) *Report on forestry mission to Ireland.* Stationery Office, Dublin.

Farrell, E.P. (1983) 'Land acquisition for forestry', in J. Blackwell and F.J. Convery (eds.) *Promise and performance: Irish environmental policies analysed.* Resource and Environmental Policy Centre, University College Dublin, pp 155–67.

Fitzpatrick, H.M. (ed.) (1966) *The forests of Ireland.* Society of Irish Foresters, Dublin.

Forest and Wildlife Service (1980) *The case for forestry.* Stationery Office, Dublin.

Gillmor, D.A. (1985) *Economic activities in the Republic of Ireland: a geographical perspective.* Gill and Macmillan, Dublin, pp 148–68.

Government of Ireland (1991) Forestry Operational Programme 1989–1993. Stationery Office, Dublin.

Hickey, D. (1990) *Forestry in Ireland: policy and practice.* An Taisce, Dublin.

Kelleher, C. (1986) 'Forestry for farmers', in *The changing CAP and its implications.* An Foras Taluntais, Dublin, pp 188–213.

McCracken, E. (1971) *The Irish woods since Tudor times.* David and Charles, Newton Abbott.

Marchand, J.P. (1975) Le reboisement dans la Rėpublique d'Irlande, *Norois* 87: 425–42.

McEvoy, T. (1979) 'Forestry', in D.A. Gillmor (ed.), *Irish resources and land use.* Institute of Public Administration, Dublin, pp 137–51.

Mollan, C. and Maloney, M. (eds.) (1991) *The right trees in the right places.* Royal Dublin Society, Dublin.

National Economic and Social Council (1979) *Irish forestry policy.* Stationery Office, Dublin.

Neeson, E. (1991) *A history of Irish forestry.* Lilliput, Dublin.

O'Carroll, N. (1984) *The forests of Ireland: history, distribution and silviculture.* Turoe, Dublin.

O'Hegarty, D. (*c.*1989) *Investing in forestry in Ireland: a practical guide.* Touche Ross and Celtic Forestry, Dublin.

# 4 Afforestation in Denmark

*Kr. Marius Jensen*

## Introduction

Denmark is located in the northern part of the deciduous forest region of Europe. The country's temperate, oceanic climate provides the natural biotope primarily for beech, but also oak and other deciduous trees such as mountain ash, lime, and European aspen are common. However, agricultural exploitation over many centuries has resulted in a nation-wide removal of forests, and today almost two-thirds of the total area is utilized for cultivation of cereals, beet crops, and grass.

The first reliable mapping of Denmark was carried out at the end of the eighteenth century. Figure 4.1 shows the distribution of forests around 1800: the eastern part of Denmark had a mixture of cultivated land and small forest areas, while the western part of the country had no woodland at all, but was covered by extensive heaths with heather and scrub. This distribution accords well with the local soil conditions. The islands of Zealand, Lolland-Falster and Funen and the whole of East Jutland have clayey soils with favourable conditions for farming, in contrast to the mainly sandy soils found in West and North Jutland. The dominant heath vegetation in these regions around 1800 was the result of unsuccessful attempts over several centuries to cultivate the soil. Such attempts, involving burning and inefficient fertilizing, gave very low yields and were usually soon followed by abandonment. Heather then invaded the abandoned land.

The background to this distribution of fertile and infertile areas in Denmark is to be found in the geological development during the several glaciations of the Pleistocene period, and, in particular, the extension of the ice during the Weichsel. Throughout this period two-thirds of Denmark was glaciated and afterwards overlain mainly by clayey till, whereas West Jutland remained ice-free; here, meltwater deposits created outwash plains around the highest-lying areas where old, leached till from Saale is still exposed.

Another factor underlying the extent of the forest area by the end of the eighteenth century was the radical land reforms which were introduced at that time. The most important of these was the enclosure movement. This completely changed the traditional cultivation pattern, whereby each farmer in the village owned small and dispersed strips of land. These were now consolidated around each farmstead, and consequently many of the old villages were split up. The opposite happened with the local forest: this had hitherto been jointly owned and exploited, but the ownership was now parcelled out to individuals, and many plots were cleared in the following years.

In the period up to 1800, many farmers got the opportunity to become owners of land. The former proprietors – the king and the nobility – lacked money and

**Figure 4.1** The distribution of the Danish forest land about 1800, covering roughly 4 per cent of the total area. © Dept. of Geography, University of Copenhagen.

therefore sold large agricultural areas with forest, for example in East Jutland. Timber and fuelwood were in great demand, and many of these new owners rapidly capitalized their woodland by felling the trees and cultivating the soil. In this way the forest area was greatly reduced.

At that time the forests supplied timber and fuelwood, cattle were grazed and pigs rooted about to find beech-nuts under the trees. The exploitation of the forests was very diversified, and the right to utilize different products was complex and shared by many people. Some local farmers might thus let their cattle graze the forest floor, others were allowed to fell trees for timber or to gather fuelwood, while the proprietor – usually a landlord – had reserved the hunting rights for himself. Consequently, a collective interest in protecting the forest against destruction did not exist, and in many cases the result was neglected woodlands, especially as far as regeneration work was concerned.

In the course of the eighteenth century, a shortage in timber and fuelwood for the towns became increasingly felt. In addition, most of the Danish fleet was lost after England's attacks on Copenhagen in 1801 and 1807. This increased the demand for the construction of new warships, but the Danish forests could not supply the required oak timber.

These warning signals as to insufficient supplies of wood products created a wide interest in rational afforestation towards the end of the eighteenth century. This interest was strongly supported by the then leader of the Danish government, C.D.F. Reventlow. Several German foresters were therefore invited to the country; for example, J. von Langen left his mark on the Crown forests, in particular on Zealand, while G.V. Brüel did good work in the first afforestation of heathland in Jutland. The increased efforts to regenerate the Crown's as well as the landlords' forests involved a specific demand to protect the areas currently carrying high forest.

The first comprehensive legislation for Danish forests was passed in 1805. In contrast to several other countries in western and northern Europe, the Danish laws on forestry were rigorous in spite of the general trend to abolish restrictions inspired by the prevalence of Adam Smith's liberal economic ideas at that time. The legislation of 1805 laid down that all Danish forests should be fenced, and grazing of cattle was forbidden. Pigs were still allowed to feed on beech-nuts as their rooting was considered good for the soil. It was further prescribed that, if cleared, all high-forest areas were to be replanted, and new owners of forest were forbidden to fell trees during the first ten years of their ownership.

There were, however, large areas with low and scattered tree growth which were not protected by the law and thus might be cleared. At the beginning of the nineteenth century the total woodland reached a minimum of about 150,000 ha, or 4 per cent of the total area of Denmark. Of this, about 100,000 ha were on the islands of Zealand, Lolland, Falster, Bornholm and Funen, while Jutland only had about 50,000 ha under forest.

## Afforestation prior to 1864

The first afforestation in Denmark took place on the heaths of Jutland towards the end of the eighteenth century. Around 1760 the Danish king established colonies for German immigrants with the objective of clearing and cultivating parts of the heath in Central Jutland. It was also attempted to plant conifers – Norway spruce and Scots pine. But knowledge about these species, and especially about their silviculture on sandy soils, was limited, and so problems and disappointments were many. The experience gained from these first afforestation attempts provided the basis for the establishment of a number of state forests in Central Jutland around the turn of the eighteenth century.

At the same time attempts were made to plant dune areas in North Zealand and along the west coast of Jutland, in order to prevent sand drifting over adjacent farmland. These plantings were supported by an act passed in 1792. In West Jutland the action was partly unsuccessful and therefore not resumed until in the middle of the nineteenth century. In dune areas, Mountain pine was usually planted, and the experience gained proved later to be very useful for forestry on the inland heath areas in Jutland.

The first afforested areas were state-owned, but soon private individuals began to be interested. In Jutland several agricultural societies encouraged their members to afforest small plots and to establish nurseries.

On the islands a number of copyholds were given up and afforested during the first decades of the nineteenth century, partly because of the country's bad economic prospects and the state bankruptcy in 1813. This meant a more compact shape of wooded areas around many larger estates.

Up to the mid-nineteenth century, however, the net result of clearing and afforestation was only the maintenance of the country's forest-covered area. Around 1850, the extent of forest was still only 100,000 ha on the islands and about 50,000 ha in Jutland – approximately the same as at the beginning of the century.

## Large-scale afforestation 1864–1914

New trends began in 1864. In that year Denmark had to surrender the two duchies of North Schleswig and Holstein after the war with Prussia. Strong nationalist movements arose, aiming to restore the nation and seeking to utilize all resources. As a result, large-scale reclamation of the Jutland heathland began, along with the first afforestation of the most infertile soils. The Danish Heath Society, founded in 1866, was an important factor, and for the following three decades was led by the dynamic engineer Enrico Dalgas.

In Jutland, the Society's objectives were supported by many estate owners, but also by wealthy persons from the islands. They invested in heath areas for afforestation and gave considerable sums to the Danish Heath Society as well as to the many societies which were established to support the local farmers wishing to afforest part of their land. The most difficult part of the process was convincing the farmers of the advantages to be achieved. The initial afforestation undertaken by the Heath Society offered good examples.

In 1884 the state decided to subsidise afforestation. The support amounted to 25 per cent of the costs involved for areas over 10 ha, subject to the conditions for forest reserves laid down in the 1805 Act. Throughout the 1890s, the state subsidies resulted in the most extensive afforestation so far.

In 1864, roughly 30 per cent of Jutland was covered by heath. The greatest extents were in western Jutland, where it was broken only by narrow strips of arable land and meadows along streams. A band of heath was also found in central Jutland, just east of the main stationary line of the last glaciation, a zone which represents the transition between the fertile eastern Jutland and the sandy soils towards the west. The landscape of this zone is varied and attractive, intersected as it is by meltwater valleys, and with great variations in soil types due to its location near a former ice margin. The cultivation of heath areas in Central Jutland began in this transitional zone. The nearby fertile land in East Jutland had gradually acquired an excess of population who had to find new livelihoods either here or, as many did, in the United States. Throughout the following 70–80 years, cultivation expanded westwards, and by the turn of the century the Jutland

**Figure 4.2** The Danish forest land in 1980, when it covered 12 per cent of the total area. © Ministry of Agriculture, Bureau of Land Data.

heathland area was halved. One quarter was now under forest and the rest was under cultivation.

During the first phase of this transformation process, almost all heath areas were cultivated in a misguided enthusiasm to utilize resources. The result was that part of the new farmland did not yield sufficiently to cover the costs of production. Over a single generation of 20–30 years, many farmers therefore gave up cultivation and undertook afforestation instead. This is the reason why so many small plantings of 3–10 ha are seen today on abandoned fields in the north–south trending transition zone through Central Jutland. Also the owners of larger plantations took in former farmland and heathland. On Djursland (the peninsula on the east coast of Jutland) extensive planting of coniferous trees was carried out in the years around the turn of this century. Djursland's undulating landscape had also been shaped by ice margins and meltwater deposits, and most of its forest is on areas of sandy soils. During the period between 1890 and 1910, 5000 ha were afforested here: half of the area was abandoned arable land and the remainder heath and small inland dunes.

Both reclamation and afforestation of heaths expanded westwards. Most of the West Jutland plantations were on heathland, and less commonly on abandoned fields of the type extensively afforested in Central Jutland. The land was more critically evaluated in West Jutland; from bitter experience it had been learned that a living could not be obtained by cultivating the poorest soils, and these were planted with Mountain pine and Red spruce. In Jutland as a whole, the forest area increased by around 100,000 ha between 1890 and 1910. About half of this increase was on the heathlands of West Jutland.

## The inter-war period

In 1912 the state subsidies for afforestation projects were reduced. This greatly slowed down planting to a level which was maintained during the inter-war period. Some special incentives were offered at a time of high unemployment around 1920 and again during the economic crisis of the 1930s. The effect of these measures can in particular be seen in West Jutland, where the increase in forest area continued, but at a slower rate than earlier in the century.

Between 1866 and 1950 the total increase in Denmark's forest-covered area amounted to 170,000 ha, including the 30,000 ha from North Schleswig in 1920. This represented more than a doubling of the area, and almost 90 per cent of the increase was in Jutland (Figure 4.2). Coniferous species comprised almost the whole of increase in area, which included the large plantations of Mountain pine established on coastal and inland dunes to protect against drifting sand.

## Afforestation policy after 1950

The development of woodlands and the composition of species throughout the last 40 years clearly illustrate the regulative mechanisms that have been involved.

Immediately after World War II, the effect was evident of the favourable subsidies introduced during the war to keep manpower away from the German occupation troops. By 1947–48 these subsidies were reduced, but the decrease in afforestation was due more to the favourable conditions for agriculture prevailing at the beginning of the 1950s. At this time, it was decreed that only the poorest soil might be afforested. Optimism among farmers was so great that some even intended to clear plantations and cultivate the soil instead.

A few years later the market for agricultural products experienced a dramatic slump. In addition, around 1960, extensive areas in Jutland suffered severely from drought and wind erosion. This led to a revival of interest in afforesting cultivated sandy soils, in particular in Central Jutland and on Djursland. In 1963, the general employment situation improved so much that subsidies for planting were greatly reduced. Over the next few years, only those afforestation projects already underway were subsidized. Throughout this period, and indeed ever since the 1890s, the funds granted by the state were administered by the Danish Heath Society.

In the mid-1960s, economic prospects for agriculture improved with the proposed entry of Denmark into the European Economic Community. A number of laws passed between 1967 and 1973 sought to protect cultivated land, and only minor areas – the poorest and least accessible – could be afforested. After Denmark joined the EEC in 1973, this outlook dominated until the second half of

**Table 4.1** The changing forest area (thousands of hectares)

| Denmark | 1861 | 1888 | 1907 | 1923* | 1931 | 1951 | 1965 | 1976 | 1982 |
|---|---|---|---|---|---|---|---|---|---|
| Hardwoods | | 151 | 139 | 148 | 149 | 140 | 147 | 137 | |
| Beech | | 109 | 108 | 100 | 104 | 91 | 84 | 75 | |
| Oak | | | 13 | 16 | 17 | 20 | 24 | 25 | |
| Conifers | | 65 | 134 | 175 | 199 | 231 | 257 | 269 | |
| Spruce & Scots pine | | | | | | 153 | 177 | 193 | |
| Mountain pine | | | | | | 56 | 37 | 29 | |
| Not planted | | 11 | 49 | 44 | 43 | 67 | 68 | 87 | |
| Total | 190 | 227 | 324 | 367 | 391 | 438 | 472 | 493 | 500 |
| Islands | 113 | 124 | 132 | 134 | 135 | 143 | 146 | 153 | |
| Hardwoods | | 94 | 90 | 92 | 92 | 84 | 85 | 80 | |
| Conifers | | 24 | 28 | 30 | 34 | 43 | 46 | 49 | |
| Jutland | 77 | 103 | 192 | 233 | 256 | 295 | 326 | 340 | |
| Hardwoods | | 56 | 49 | 55 | 57 | 56 | 62 | 57 | |
| Conifers | | 41 | 106 | 145 | 165 | 188 | 211 | 220 | |

*Including North Schleswig

the 1980s. In agriculture, there was an optimism during these years, and the keyword was increasing production.

Through more than a decade the wish to abandon farmland in favour of traditional forestry was therefore absent; other potential areas, such as heaths and bogs, had gradually been reduced and what was left had to be protected against both afforestation and cultivation. At the same time, however, a new market niche was discovered in forestry and agriculture in the 1950s and 1960s in the form of Christmas trees and greenery. By law, these products were exempted from the restrictions imposed on alternative use of agricultural land. Species such as the firs *Abies nordmanniana* and *Abies nobilis*, and to a lesser extent also Norway spruce, were planted as Christmas trees. Throughout the 1970s and 1980s, about 1500 ha were used in this way, or 3 per cent of the Danish woodland area. From an economic viewpoint, however, it has a much greater importance, as it now represents well over one-third of the total production value.

## Forest species and ownership

The expansion of Danish wooded areas from 150,000 ha in the middle of nineteenth century to over 500,000 ha by 1980 is summarized in Table 4.1 and Figure 4.3. Of the 500,000 ha, however, only about 80 per cent is under trees, and there is a clear trend towards more open areas in forests to serve recreational purposes. The latest nation-wide count of the forests is from 1976, but the trends established between 1965 and 1976 have continued until the present.

During these years the islands had a small increase, which is exclusively due to planting of coniferous trees. The area under deciduous trees has been almost constant, with a slight decrease during the most recent decades. This is especially

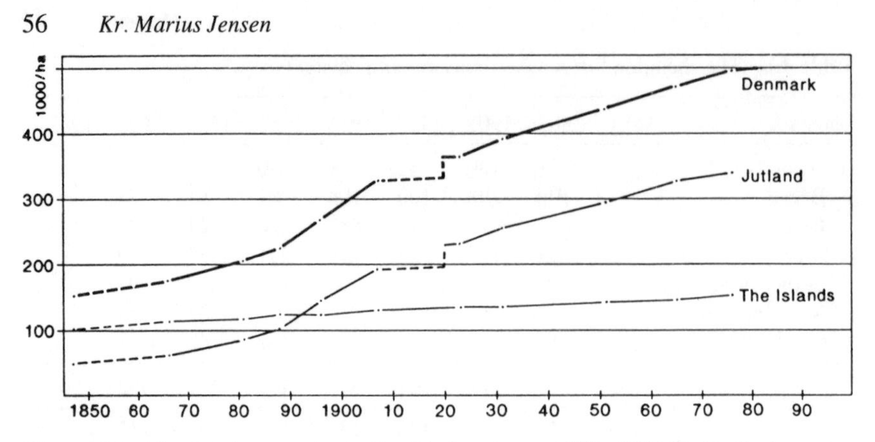


**Figure 4.3** The development in the Danish forest area, 1850–1980. By far the largest increase has taken place in Jutland. The breaks on the curves for 1920 stem from the incorporation of North Schleswig.

seen for beech, whereas oak has become more common. The shorter rotation length of 50–60 years for coniferous trees, compared with 100–120 years for deciduous trees, gives the former a clear economic advantage.

Expansion has been far more dramatic in Jutland. Here, the wooded area has increased sevenfold (including the 30,000 ha from North Schleswig), but again coniferous species account for the increase. It should be noted, however, that around 15 per cent of Jutland's woods are non-commercial, as they have been planted on dunes with the main objective of preventing sand drift.

The Danish woodland area is distributed over many ownership units. In 1976 they numbered 26,000, but amalgamations are continuing. Small units of 0.5–5 ha dominate numerically, accounting for 70–80 per cent of the total, but they comprise only 7 per cent of the area. On the other hand the 145 units of more than 500 ha account for as much as 55 per cent of the area (Table 4.2).

Table 4.3 sets out the patterns of ownership and composition. In contrast to other sectors in Denmark, central and local government bodies own a significant part (about one-third) of the forest area. The state forests are concentrated in a few large units of over 1000 ha. They are mainly found in old forest areas and on sand dunes. They are distinctive in the extent of their open areas: because state forests are used for recreation. In contrast, the private forests are small and are widely distributed over the whole country.

## Recent and future forest policy

During the 1980s, the EC countries had a surplus of agricultural produce, and the high yields of cereals in particular posed problems for the EC economy. It was therefore agreed by the EC and the national governments in the member countries to subsidize fallow land or afforestation in order to reduce cereal production. In Denmark policies of 'set-aside' (fallowing) have not been successful, as traditionally most farmers have the attitude that even poor soils should be cultivated. Few have therefore been tempted by the incentives to give up cultivation. Plans

**Table 4.2** Distribution of forest units by size, 1976

| | Total | 0.5–5 ha | 5–10 | 10–50 | 50–100 | 100–500 | 500 –1000 | >1000 ha |
|---|---|---|---|---|---|---|---|---|
| Denmark | 26,000 | 18,850 | 3,300 | 3,000 | 300 | 380 | 75 | 70 |
| Islands | 5,000 | 3,970 | 430 | 375 | 75 | 115 | 35 | 35 |
| Jutland | 21,000 | 14,880 | 2,870 | 2,625 | 225 | 265 | 40 | 35 |

**Table 4.3** Forests by ownership and species, 1976 (thousands of hectares)

| | Hardwoods | Conifers | Open areas | Total | % |
|---|---|---|---|---|---|
| State and municipalities | 30.9 | 89.9 | 49.2 | 170 | 34.4 |
| Danish Heath Society | 0.5 | 9.5 | 2.1 | 12 | 2.5 |
| Foundations | 11.8 | 11.2 | 4.1 | 27 | 5.5 |
| Companies | 12.3 | 37.8 | 8.6 | 59 | 11.9 |
| Private ownership | 81.5 | 121.0 | 22.9 | 225 | 45.7 |
| Total | 137 | 269 | 87 | 493 | 100 |

for extensive afforestation were first introduced in 1989 when a number of Acts were passed in the Danish Folketing. Farmers might again undertake afforestation, but in accordance with regional plans.

Each county was required to submit plans during 1990 and 1991 and to indicate those areas which might be suitable for afforestation and at the same time those areas where new forests would not be acceptable. The intention of these measures is to protect unique and valued landscapes where afforestation might impair the scenery and obstruct its attractive views. Together, the suitable and restricted areas amount to only about 20 per cent of the total farmland area: the remaining 80 per cent is not affected by either measure.

The new afforestation plans are designed with the objective of doubling the Danish forest area over the next 75 years. At present, the country imports about two-thirds of its requirements of forest products. A doubling of the forest area would increase the production of timber, increase the recreational potential, and create more scope for the wild fauna and flora. In order to maximize these benefits, part of the proposed forest area is planned to be established near major towns, and the rest on sandy soils and in regions of difficult access.

To promote afforestation, provision has been made for state acquisition of land, and private initiatives are subsidized. The largest grants are given for schemes on land specifically identified for afforestation, but projects on other areas may also be supported. On the other hand, no subsidies are normally granted on areas where planting is considered undesirable. The planting of Christmas trees and species for greenery is still possible in all areas, but will not be subsidized. The choice of type of planting and of species is at the discretion of the afforester, but the planting of broadleaved species is encouraged by higher levels of subsidy.

The proposals to expand the forest area have been received with mixed feelings by farmers. Even though the whole policy is based on voluntary principles, scepticism about the regional plans is evident. The farmers fear restrictions on their traditional rights of disposing freely of their own land, and claim that the price levels of farms might be influenced. The fear is probably exaggerated, as only small amounts have been reserved so far for the extension of the Danish forest area. The attitudes of environmentalists towards further expansion of forest are generally very positive, especially where broadleaved woods with open areas are being established.

The surplus production of cereals in the EC and the considerable reduction in prices for farm products expected in coming years will marginalize much hitherto cultivated land in Denmark, as in other parts of Europe. For the individual farmer, however, it is a difficult decision to stop cultivating his land and to afforest it. A large investment and no profit for many years will naturally discourage many farmers. At the same time the employment situation is worsening in rural districts. People leave and this again has a negative effect on the infrastructure. In return, many landscapes become more attractive to the public, and reduced cultivation should also mean less pollution to water courses and lakes.

There are thus comprehensive and complicated problems confronting the intention to double the country's forest areas by the year 2050, but the trend is evident. Denmark is going to be still more wooded, although the rate will probably not keep space with the 'Green Plan' outlined by the politicians.

## References

Henriksen, H.A. (1988) *Skoven og dens dyrkning*. Nyt Nordisk Forlag, Copenhagen.

Jensen, Kr.M. (1976) *Abandoned and afforested farm areas in Central Jutland. Atlas of Denmark II, 1.* The Royal Danish Geographical Society, Copenhagen (in Danish with summary and figures in English).

Reventlow, C.D.F. (1879) *A treatise on forestry 1879*. (Published by the Society of Forest History, Copenhagen, 1960.)

# 5 Hungary

*B. Keresztesi*

In the course of her 1100 years' history, the frontiers of Hungary have changed several times. The so-called 'historical Hungary', the Kingdom of Hungary, comprised the entire Carpathian Basin, a geographical macro-region in Central Europe with favourable climate and flanked by the Carpathian Mountains.

The present-day frontiers of Hungary were fixed by the Treaty of Trianon signed in 1920, just after World War I. This treaty ordained the disintegration of the multi-national country by the secession of Slovakia, the Carpathian Ukraine, Transylvania and Croatia, transferring large territories and large Hungarian populations to Czechoslovakia, Romania and Yugoslavia. As a consequence of the Treaty of Trianon, historical Hungary lost 67 per cent of her territory and 84 per cent of her forests, as is well illustrated by Figure 5.1.

Most of the mountainous regions covered by forest (the Carpathians) were lost, and only the lowland and hilly regions encircled by the adjoining mountains, mainly used for agriculture, remained in Hungary.

In this chapter, the subsequent information and data refer to the present-day territory of the country.

## Trends in the forest area over the centuries

The natural environment of present-day Hungary at the time of the Conquest (AD 896–900) was summarised by Sándor Somogyi (1988) as follows:

> Reconstructions of the way of life of the conquering Hungarians show that they came to our country as livestock-raising and shepherding people cultivating the land around their winter lodges. The dominance of livestock-raising and shepherding alternating with agricultural cultivation of suitable lands had been a characteristic feature of the people living in the lowlands and hilly regions of the Carpathian Basin for several millennia. Thus, social impacts on the natural environment had to a great extent already manifested themselves by the time of Conquest. An early human impact was, for example, that which impeded the natural reforestation of the Great Plain during the Atlantic and Sub-boreal phases some millennia ago. With this spatially expanding and increasingly complex socio-ecological influence, it is obvious that the initial natural environment had suffered considerable changes by historical times.

The present-day environment is the cumulative product of these and subsequent changes arising from human impacts over the centuries.

At the time of the Conquest, Hungary's forests may have covered an area of about 3.45 million ha. It was King Joseph II who ordered that the forests be

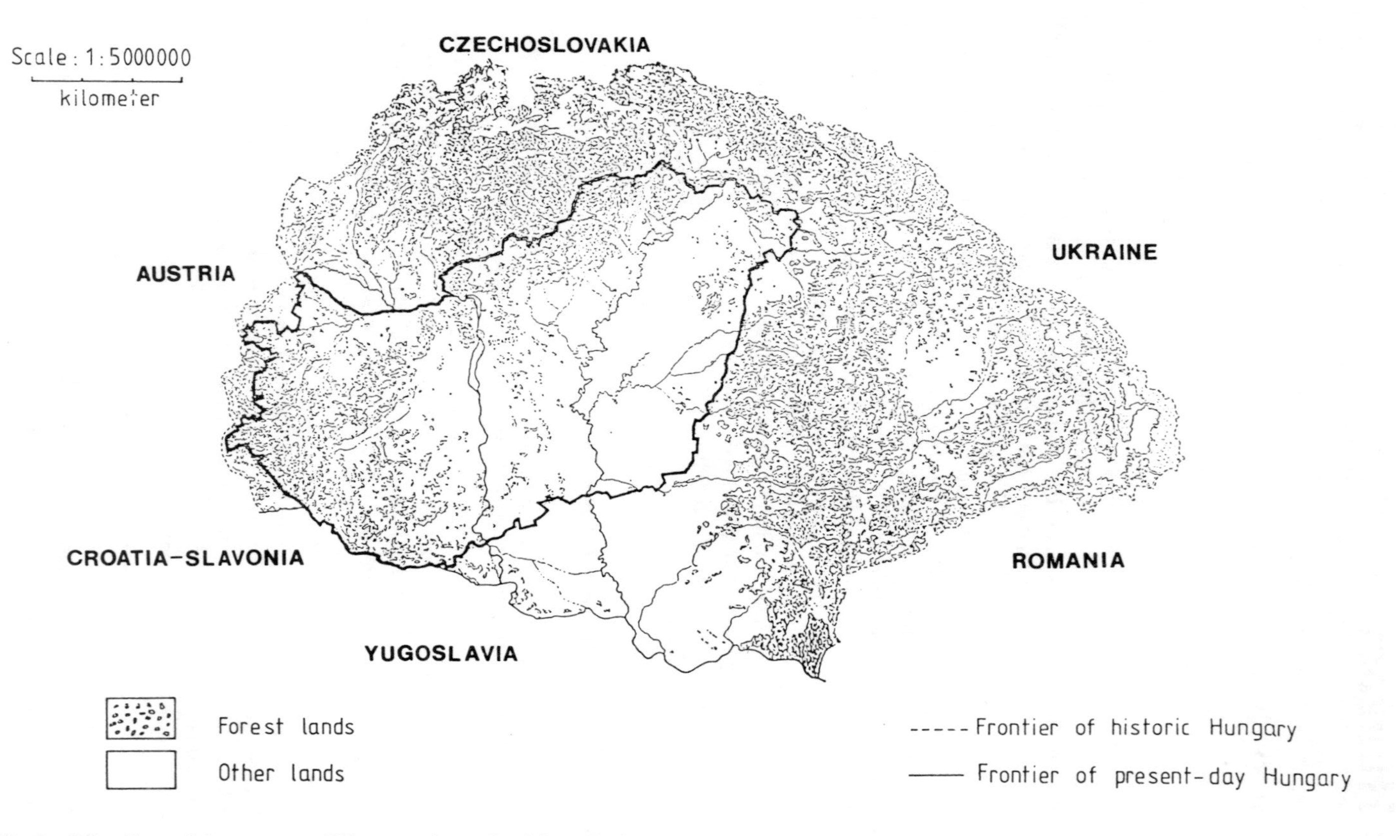


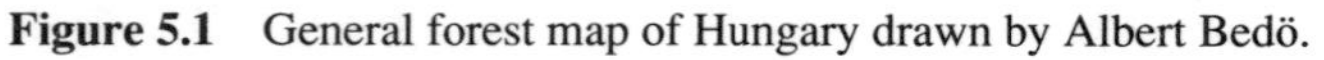

**Figure 5.1** General forest map of Hungary drawn by Albert Bedö.

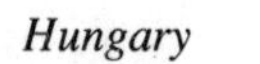

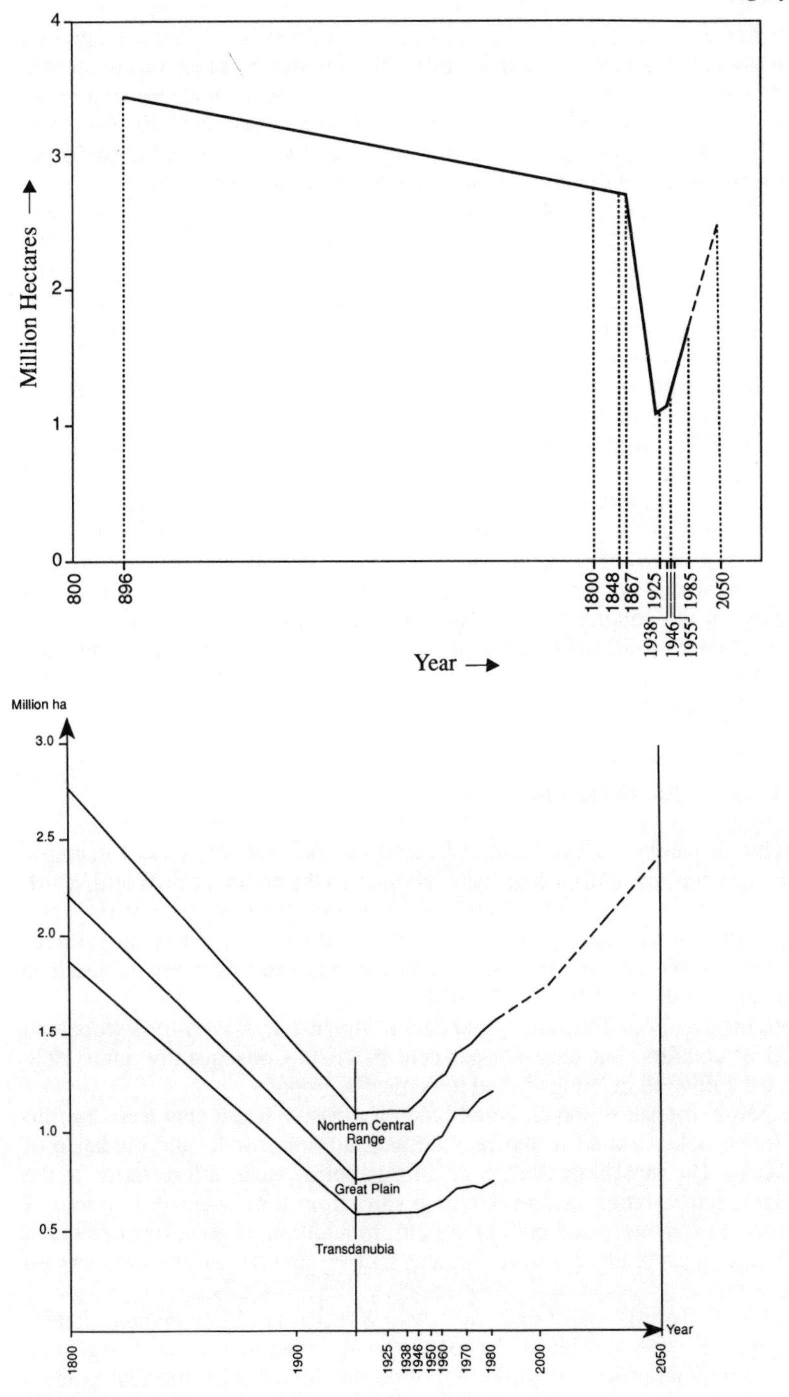


**Figure 5.2 a and b** Changes in forest area by regions.

mapped. According to the resulting maps, 2.766 million ha of forest existed on the present-day territory of the country in 1800. This suggests that during the 900 years between the Conquest and 1800, only 0.684 million ha or 20 per cent of the total initial forest area had been deforested. During the subsequent 120 years from 1800 to the Treaty of Trianon, 1.675 million ha of forest, amounting to 50 per cent of the woodland at the time of the Conquest, was lost (Figure 5.2).

Until the Revolution of 1848, most of the forests belonged to the landowners. Peasants had the right to collect fuelwood and timber and pick nuts for their own needs, graze their livestock and feed swine on mast in the landowners' forests. After the Revolution, the peasants were liberated and to compensate them for these rights some forests were given to them from the landowners' holdings. They were able to dispose of these freely.

After the Compromise of 1867, a period of vigorous economic development began in the country, resulting in economic and social transformation over the next few decades. Large-scale railway construction, the development of industry and mining, and the growth of cities all needed a large amount of wood and capital. As a result, extensive forest areas were exploited and deforested by landowners, as well as by peasants, over a period of about 50 years. As a consequence of these activities, in addition to flood control and water regulation measures, the natural environment changed rapidly and very unfavourably. In this period parts of the country, which until then had suffered little human impact, were brought into cultivation. Our present-day ecological and environmental problems result mainly from the rapid and massive deforestation which occurred between 1867 and 1925.

## Afforestation of the Great Plain

The origins of planned development of the Hungarian forestry (forest management, primary timber industry and trade) go back to the period after World War I. The forests in the lowland and hilly regions had only slight importance for wood supply prior to the war. On the large forest estates, the landowners in the present-day territory of the country pursued coppice management systems to produce fuelwood and small-sized industrial wood.

Before the Treaty of Trianon, 24 per cent of the historical country's woodland consisted of conifers, but only 4.1 per cent in 1920. Consequently, after 1920 demand for softwood in Hungary had to be met by imports.

In response, foresters, and especially the progressive forest engineers, formulated a forest policy suited to the prevailing conditions, under the guidance of Károly Kaán. The most important goals of this policy were: afforestation in the Great Plain; reforestation of bare lands; a shift from a management system of clear felling to a shelterwood cutting system. In addition, it was proposed that a legal obligation of planned forest management be applied to privately owned forests.

The idea of afforestation of the Great Plain was first raised by two physicians, János György Krámer and János Gregory, who observed that forests and plantations had beneficial effects on the health of people. In 1739 Krámer published a book on 'Morbus Hungaricus', in an appendix to which he discussed the harmful effects of the lack of forests and trees in the Great Plain, and pointed out the great importance which the afforestation of the Great Plain could have in respect of public health.

From then on, afforestation of the Great Plain was, in a sense, permanently on the agenda, but the government in power did not consider it to be an urgent task. It was not until 1923 that Károly Kaán, the Secretary of State in charge of forestry administration between 1916 and 1923, managed to have the Law on the Afforestation of the Great Plain enacted. According to this law:

> For the promotion of agricultural production in the Great Plain, furthermore for the improvement of health and climate conditions, as well as for meeting the demand of the population for timber forests and groups of trees, windbreaks or rows of trees should be established in a planned distribution in the territory of the Great Plain's municipalities.

The law ordered that quicksand areas, flood plains and alkaline soils should primarily be afforested. It was planned to establish 109,340 ha of forests and plantations, of which 51,555 ha were afforested by 1938, mainly in the Danube–Tisza interfluve and in the Nyírség region. Of the new plantation area, black locust accounted for 72 per cent of the area, oaks 2 per cent, other broadleaved species 22 per cent, and conifers 2 per cent. Taking into account the deforestation carried out in the meantime, the net increase in the forest area was only 18,456 ha.

Kaán's role as chief of forestry administration was governed by his firm conviction that 'forests and their management play such an important role in the national economy that they have to be judged and valued always from the higher point of view of the national economy'. His programme on forestry development reflecting this conviction was summarised in his book, *Forestry policy matters*, published in 1923. In this programme, he considered the replacement of the former Forest Law (Act XXXI of 1879) by a new law to be the most important task to ensure the maintenance of forests and effective development of forestry.

It is a great pity for Hungarian forestry that, because of the power relations of that time, he did not manage to table a new Bill on the 'maintenance of forests'. In the draft Bill he wanted, as he said, to put an end to the misuse and misinterpretation of personal rights, freedom in enterprise and other economic action and other attractive slogans in the field of forestry, where freedom without responsibility can lead to the wastage of accumulated resources and involve serious, far-reaching consequences. In his view, therefore, it was in the interests of the national economy that rational use of forest resources be ensured, if necessary by state intervention and legal regulations limiting personal rights in the field of forestry. This was the only way of preventing overcutting and felling without regeneration in privately owned forests.

Kaán considered the rejection of his draft Bill as a deliberate obstruction of his most important effort, and as a consequence he retired in 1925. He later refused to accept the new Forest Law that was enacted as Bill IV of 1935, thinking it no more than a poor compromise.

Though the main objective of Hungarian forestry policy was to eliminate the deficit of the timber trade (i.e. by permitting a greater degree of self-sufficiency in timber), the issue of multiple use of forests was also raised in the Law of 1935. The main aspects of this issue were summarized in the preamble to the Law:

The multi-faceted role which forests play in wood supply on the one hand and in protecting the soil, in controlling and balancing the powers of nature, and in improving public health on the other hand, assigns such a high importance to forests that forest maintenance and forestry itself cannot be allowed to be appreciated and judged merely from the viewpoint of the private owners' rights, especially since the protective impacts of a forest affect also the lands of others. Thus, the maintenance of forests is required by the public interest attached to the forests' economic, public-security, public-health and nature-conservation functions, especially since forests can fulfil their protective function only when their soil cover and their stands are in good condition.

## The national afforestation programme

The conditions for establishing modern forestry in Hungary were created after World War II. The nationalisation of forests and timber-industry plants, as well as the introduction of centrally planned management for the whole economy, enabled intensive forestry and modern forest industries to be established, keeping future interests also in mind.

The new political leaders of the country confronted the forestry industry, urging a marked reduction in felling to compensate for wartime overcutting. The situation improved significantly by the end of the 1940s, when the government provided in the First Five-Year Plan (for 1950–1954) previously unimaginable financial support for the reforestation of the areas cut during the war, and for afforestation. As a result, 110,000 ha unsuitable for agricultural use were afforested.

András Hegedús, the Minister of Agriculture, in 1954 put emphasis on professional guidance and, therefore, established a new senior management structure and appointed professional experts to the new posts. The Chief Forest Engineer was put in charge of elaborating general ideas on the development of forestry and the timber industry, on the basis of the research achievements and practical experience accumulated in Hungary and abroad. The Council of Ministers approved the elaborated and widely discussed plans, and their conclusions were enacted in Government Decree No. 1040/1954 on the development of forestry, and in Decree No. 3009/1955 on the development of the forest industries and on the measures for wood saving.

At that time forestry in Hungary employed 1,000 well trained forest engineers and more that 3,700 forest technicians, which even compared with world standards was a very good supply of professionals. This professional capability, in combination with the opportunities and advantages provided by land reform, the nationalisation of the timber industries and the centralised direction of the economy, allowed the successful development of forestry.

Kaán's afforestation programme for the Great Plain was broadened into a national programme including ecology. Ecological conditions at the turn of the eithteenth–nineteenth centuries were still relatively sound, but then deteriorated rapidly up to the time of World War II as the consequence of enormous deforestation and large-scale water-regulation works. In the first half of the 1950s, the trends were reversed through the creation of large new forests and

**Table 5.1** Forest area by regions, 1800–1985 (thousands of hectares)

| Year | Transdanubia | Great Plain | Northern Central Range | Total |
|---|---|---|---|---|
| 1800 | 1878.7 | 329.5 | 557.6 | 2765.8 |
| 1925 | 591.6 | 180.6 | 318.6 | 1090.8 |
| 1938 | 598.6 | 189.1 | 318.3 | 1106.0 |
| 1946 | 607.7 | 199.1 | 317.4 | 1124.2 |
| 1950 | 619.9 | 245.0 | 301.0 | 1165.9 |
| 1955 | 644.0 | 315.3 | 298.1 | 1257.4 |
| 1960 | 665.4 | 343.0 | 297.8 | 1306.2 |
| 1965 | 729.9 | 376.8 | 314.8 | 1421.5 |
| 1970 | 745.5 | 366.6 | 358.6 | 1470.7 |
| 1975 | 743.6 | 438.0 | 363.7 | 1545.3 |
| 1980 | 769.8 | 466.5 | 373.9 | 1610.2 |
| 1985 | 775.3 | 490.7 | 381.9 | 1647.9 |

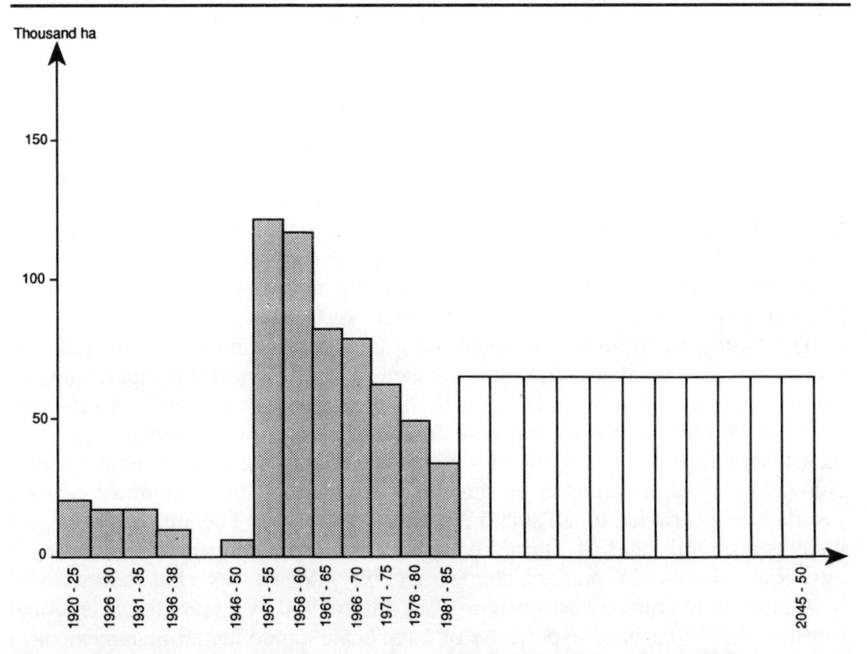


**Figure 5.3** Afforestation in Hungary since 1920 (projected rates from 1985).

shelterbelts and the build-up of irrigation systems. The aim now was to re-create the situation as it existed in the time of King Joseph II.

The national afforestation programme sought to realise goals of soil and land protection, water conservation and benefits in tourism, hunting and fishing, through the creation of a system of forest and shelterbelt plantations all over the country. The afforestation was integrated with other forms of land use and management (farmland, shelterbelts, water reservoirs, irrigated lands) to create an organic system of optimal landscape structure. The results achieved are shown in Table 5.1 and Figure 5.3.

The total forest area increased between 1925 and 1938 by 15,200 ha and by 523,700 ha between 1946 and 1985. The increase in Transdanubia in the two periods was 7,000 and 167,600 ha, on the Great Plain 8,500 and 291,600 ha, while in the Northern Central Range the forest area decreased by 300 ha and increased by 64,500 ha, respectively, in the two periods. A fundamental change took place in the Great Plain. There the forest area increased during the inter-war period by 4.7 per cent and in the last 40 years by 146.5 per cent. Between 1946 and 1985 about half the abandoned agricultural area (farmlands, vineyards, pastures and orchards) was afforested. Of the area afforested, 148,000 ha were planted by state forest enterprises, 97,000 ha by state farms, 154,000 ha by co-operatives and 125,000 ha by other state agencies.

The modernisation and intensification of large-scale farming freed more and more land which was no longer suitable for profitable farming, but was still suitable for afforestation. The afforestation of such land continues.

Afforestation was especially rapid between 1947 and 1954, when 293,000 ha of land were regenerated or afforested. It was found that 2.4 ha had to be planted to get one hectare of established regeneration and afforestation, i.e. only 42 per cent of the planted trees survived. Thus regeneration and afforestation had to be much improved. The aim was to plant only 1.6 ha for one hectare of established forest, i.e. to achieve a survival of 62 per cent. To this end, site assessment was intensified, and mechanical stump grubbing and deep ploughing after final cutting were introduced. On the basis of the species composition targeted for the end of the 1950s, nursery production was prescribed for the state forest enterprises five years ahead. Regular tending of the regenerated and afforested areas was made compulsory, and three forest protection stations were set up to control forest damage. These measures were generally successful.

The basic precondition for determining the species composition of the new forests was the fact that Hungary is situated in the Central European deciduous forest zone. Accordingly, an increase in the area of oak forests and a decrease in the area of Turkey oak stands were foreseen. Fast-growing poplar and black locust were to be planted on all sites where site conditions were suitable for their cultivation. Conifers were to be planted if silvicultural or economical reasons justified their introduction. Table 5.2 outlines the results. The area of beech was maintained, while that of Turkey oak and hornbeam declined. The substantial territorial gain of oaks, poplars, conifers and black locust was a big success.

A new silvicultural system was worked out in 1956 by a team of professional foresters on the basis of experience of large-scale experimental management in the estate of Sárvár-Farkaserdő with the aim of growing large, valuable logs in a shorter time, boosting the value of final cuttings, and in the long run establishing better, more valuable forest stands.

Until World War II there was no regular and appropriate silvicultural management in private forests. In relation to the principles of intermediate cutting, the ideas of the old German school prevailed ('low thinning'), which meant the cautious removal of suppressed, malformed or unhealthy trees, while maintaining crown closure.

**Table 5.2** Changes in species, 1948–1985 (thousands of hectares)

| | Species Stocked forest area | | Change | |
|---|---|---|---|---|
| | 1948 | 1985 | Increase | Decrease |
| Oaks | 283.2 | 345.9 | 62.7 | – |
| Turkey oak | 191.9 | 176.7 | – | 15.2 |
| Beech | 101.3 | 101.2 | – | 0.1 |
| Hornbeam | 102.3 | 96.9 | – | 5.4 |
| Black locust | 199.1 | 268.8 | 69.7 | – |
| Poplars | 33.8 | 149.7 | 115.9 | – |
| Other deciduous spp. | 87.6 | 124.8 | 37.2 | – |
| Conifers | 67.5 | 229.1 | 161.6 | – |
| Total | 1066.7 | 1493.1 | 447.1 | 20.7 |

The requirement of selecting and marking the best trees of the stands to be maintained up to the final cutting brought a new viewpoint in silviculture – active, positive selection work. All thinnings are carried out primarily in the interest of these trees. The new silvicultural system resulted in a significant increase in the volume of thinnings. The best result was registered in 1960, when 43.5 per cent of the total fellings were obtained from thinnings. The greatest volume (27.0 per cent) of shelter wood cuttings was registered in 1970.

Frequent and heavy game damage is an unsolved problem of silviculture, which has gone unnoticed in the professional and financial plans and balances in forestry. Forest damage caused by game was not even precisely measured. It is almost impossible to detect changes in forest stands, in the course of the typical forest rotation of many decades, with the aid of regular routine forest inventories (every ten years). The damage caused by weather, epidemics, game and irresponsible silvicultural work does not appear in the balances and is manifest as production losses. Prolonged game damage (chewing, breaking, brushing, bark peeling), however, leads to losses in yield and quality, making silvicultural investments futile. It is therefore absolutely necessary to co-ordinate short- and long-term planning in forestry and game management, to draw up inventories precisely on game damage and to determine the level of damage tolerance by forest types and to regulate the big-game population density.

Progress in Hungary in the past three decades can be summarised as follows. The forest area of the country rose from 1.1 to 1.6 million ha. Growing stock increased from 165 to 274 million $m^3$, and allowable cut from 3.1 to 8.3 million $m^3$. These achievements made it possible, in line with an increase in wood consumption from 5.7 million $m^3$ (equivalent) to 9.5 million $m^3$, for the percentage of domestic wood production in relation to total wood consumption to rise from 54.7 per cent in 1950 to 70.5 per cent by 1985. Timber exports amounted to 110,000 $m^3$ (equivalent) in 1955, but 2,035,000 $m^3$ by 1985. Further details of the results achieved are shown in Keresztesi (1991).

## The great national green programme

Nowadays the errors of the recent past and their possible correction are the main topic of everyday conversation in Hungary. Nevertheless, long-term programmes are required in forestry and in other sectors of the economy.

In 1988 a prognosis was made for agriculture by Pál Izinger and András Madas. They found that average yields in plant production might by AD 2000 achieve those of the most advanced large-scale farms in 1986. This may result in the abandonment of 1 million ha of agricultural land, of which 850,000 ha, primarily abandoned pastures, could be considered for afforestation. Based on this prognosis, a 'great green' programme consisting of two schemes of afforestation was elaborated. One involved, as an initial phase of the whole programme, the afforestation of 150,000 ha in the years 1991–2000. This project has been approved by the government and financed from year to year by the state budget. The other was a long-term plan prepared jointly by the former Ministry of Agriculture and Food and the Hungarian Academy of Sciences. This project set as an aim for the period up to the year 2050 the afforestation of 700,000 ha of land unsuitable for profitable agricultural production.

This 'great national green programme' accords well with the great international movement which considers the protection and development of the environment as the most important global task that can no longer be deferred. In this respect forestry, including both the protection of existing forests and the establishment of new forests, plays a prominent role. As M. Pavan (Italian professor and member of the Club of Rome) said in 1972:

> Ruthless exploitation of forests is a very convenient management practice. It results, however, in the degradation of soil, which has to be reclaimed and reforested. It is therefore preferable to agree on a global scale on a reasonable forest extent that has to be preserved. If mankind does really care for its future, then a global action plan should be formulated for the rehabilitation and preservation of the environment and use ecological principles so as to make it possible for mankind to live on our globe for ever.

The 'great national green programme' will result in an increase in the forest area to 2.5 million ha. This, together with the area of orchards and vineyards (186,000 ha) will give a total of 2.7 million ha, which was the forest area of the country at the time of Joseph II. Thus, when new forests are established in such a way that the environment can be improved, then the relatively favourable ecological conditions that prevailed two hundred years ago will be largely restored.

In Hungary, the ability of trees and forests to absorb carbon dioxide might be of crucial importance in efforts to abate the greenhouse effect. As a result of the latter, the frequency and strength of droughts is increasing, and certain maize-growing regions might shift from the 'possible' to the 'essential' irrigation zone. This may affect also intensively grown poplar, willow and black locust varieties. A basic objective is to improve the balance of emission and absorption of carbon dioxide.

In the framework of the 'great green programme' more than half of the new forest planting will be with native oaks and other deciduous species. In this way, the reconstruction of quasi-natural forests will be sought. At the same time, special-purpose forests will also be established: 30,000 ha for producing veneer logs of hardwood broadleaved and poplar species; 140,000 ha of high-yielding coniferous plantations; 65,000 ha of fodder forests to produce forage for domestic animals and big game; 60,000 ha of black locust forests of late-flowering varieties to increase commercial honey production; 40,000 ha of energy forests primarily to supply village populations with fuelwood – in total, 335,000 ha of new forests.

The implementation of the 'great national green programme' is thus of historical importance in terms of environmental improvement. At the same time it is an attempt to make the country self-sufficient in wood products through an increase in export earnings to compensate for the costs of imports. In addition, it facilitates the alternative use of land on co-operative, state and private farms that is unsuitable for efficient agricultural use.

The securing of popular support for this programme is a major task. In this respect the statement made in 1927 by Károly Kaán, who initiated the afforestation programme of the Great Hungarian Plain in the 1920s, deserves special attention:

> It is necessary that every Hungarian citizen be aware of the great importance of this issue that will increase the strength of the nation and the national economy. The government, in turn, after overcoming the legal difficulties and launching adequate grant aid should encourage and support the population in implementing this well-intentioned programme.

It is a firm conviction of foresters that it is the maintenance of existing forests and the establishment of new forests which play the most important role in improving the environment. For a relatively small capital investment, the human environment can be widely improved. Well-organised afforestation can be an important contributor to the development of the countryside, as well as to the reduction of unemployment. In the course of implementing the Law on Compensation it will be impossible to satisfy all claims to good arable land, but on the other hand probably not all the claimants intend to use the land for agriculture. Therefore, it would be advisable to offer financial support to those who are willing to establish forests, primarily for supplying fuelwood for the rural population (energy forests), for producing forage for livestock (fodder forests), or especially for creating bee pastures. It is in the national interest to get the support of the rural population for this programme.

## Forest resource management

For forestry (silviculture and logging) there is no alternative other than planned management. Managing the forests according to long-term plans is the only way of accomplishing sustainable forest management, keeping in mind the interests of future generations.

In the past four decades Hungarian forestry has developed in line with the national economy and has tried to adapt itself to world trends in forestry. In the 1950s and 1960s, under the system of central planning of the economy, considerable achievements compared even with international standards had been attained in increasing the production of forest resources, thus creating the basis for felling

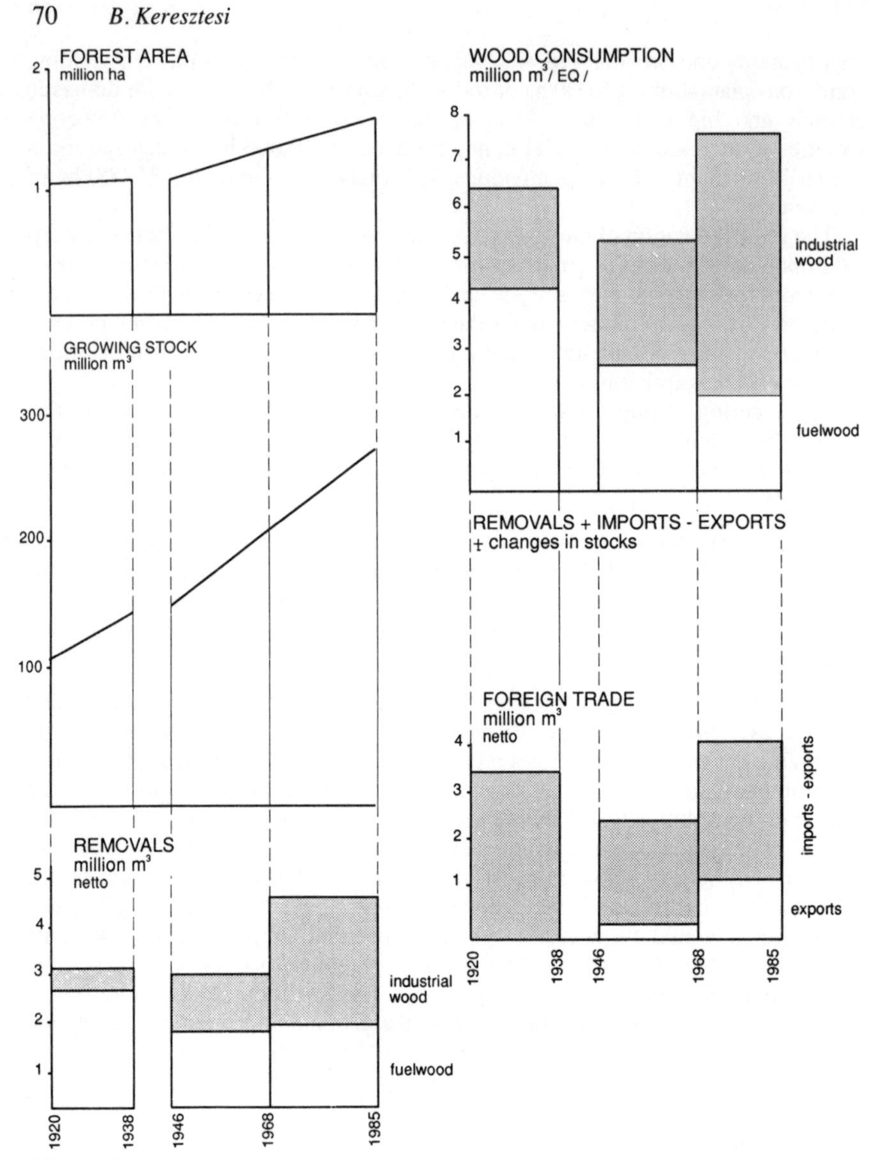


**Figure 5.4** The development of forestry in Hungary between 1920 and 1985.

to be increased subsequently.

In 1968 economic reforms, which sought to simulate market mechanisms by planned regulation of economic processes, were applied also to forestry. The new system was based on profit motive and incentives and enabled forestry, on the basis of results achieved in the production of forest resources in the preceding decades, to make a considerable contribution to easing the problems of the national economy by improving the foreign trade balance and the equilibrium of the budget in the period of the 1970s economic crisis.

At the same time, however, serious problems arose, because the new system neglected long-term interests and responsibility, which are the basic requirements in forestry. Consequently some enterprises concentrated felling in the best stands, without carrying out the reforestation of cutting areas and the necessary stand treatments. The situation became even worse because silviculture and game management in forests could not be properly co-ordinated. Overpopulated game stock made the regeneration of forests increasingly impossible. As a result, the actual increase in growing stock between 1981 and 1985 was 1.5 – 2.0 million $m^3$ less than it would otherwise have been. The reform created an adequate profit orientation for harvesting and wood processing, but it did not encourage the reforestation of cutting areas and the expansion of forest resources, which are the basis of long-term development in forestry. In other words, forestry accounting needs to look beyond annual profit and export. To this end, forest-resource balances are to be prepared every five years. If this were done retrospectively over the last 65 years, enlarged reproduction would emerge as the main trend (Figure 5.4). Progress was modest between the wars, but was outstanding between World War II and the economic reform of 1968. In the long term, self-sufficiency can be achieved with the value of softwood imports being covered by that of hardwood exports. The costs of reproduction (regeneration, stand treatment and forest protection) have since 1956 been covered by the Reforestation Fund out of the sales revenue of the forest management organisations. Afforestation has been financed entirely from the state budget as a state investment.

## References

Kaán, Károly (1923) *Erdogazdaságpolitikai kérdések.* Rötting-Romwalter Nyomda Részvénytársaság, 102 old. II. kiadás.

Somogyi, Sándor (1988) A magyar honfoglalás földrajzi környezete, *Magyar Tudomány* 88/11. szám 863-869. old.

Keresztesi, B. (ed.) (1991) *Forestry in Hungary 1920–1985*. Budapest, Akademiai Kiado.

# 6 Afforestation policy and practice in Spain

*Helen J. Groome*

## Introduction

The concept of afforestation, or the deliberate planting of trees as opposed to their natural regeneration, only made its appearance in Spain when deforestation had reached economically or socially unacceptable limits. Thus, although present to some extent in earlier centuries, awareness of deforestation became critical when forest products, above all firewood, charcoal and timber, became increasingly scarce, from the eighteenth century onwards (Bauer 1980; Urteaga 1987). A variety of human processes contributed to this loss of forest and woodland, amongst which the most important were:

- growing demographic pressures, which increased demand for household forest products (above all firewood and timber) and led to over-exploitation in those woodlands nearest to settlements; forests were also cleared and transformed into arable and pasture lands.
- over-grazing (although the role of the Mesta[1] in deforestation is still a highly contentious point of debate).
- incessant wars, which employed scorched-earth techniques, created a high demand for timber for shipbuilding and consumed huge amounts of wood in battles.
- growing overseas commercial relations which increased the timber demand for merchant shipping.
- important and growing industrial activities which increased wood-based energy demands (mainly for charcoal) and manufacturing demand (pitprops for mines, timber for components, etc.).

A number of natural phenomena also contributed to deforestation, perhaps the most important being the various diseases that seriously affected certain tree species such as the sweet chestnut (*Castanea sativa*, affected by *Phytophthore cambivora* or *P. cinamomi*) and oaks in general (affected by *Oidium* ssp.), above all in the northern Atlantic part of Spain.

Until the nineteenth century, no rational, orderly or strictly enforced forest policy existed to cope with deforestation and its social, physical and economic consequences. Neither, it should be added, had a co-ordinated policy been developed to guarantee the maintenance and natural regeneration of the remaining woodland.

After centuries of specific, localised forest legislation, based overwhelmingly on penalising bad management practices or crime, and only rarely on the promotion of positive techniques of woodland conservation, regeneration and utilisation, the nineteenth century saw the confluence of a series of factors that made the establishment of a forest policy possible, and with it, an incipient afforestation policy. Apart from the growing awareness of deforestation, and its perception as a clearly negative process when exceeding certain physical limits, the introduction of agricultural and silvicultural sciences (the latter based mainly on French and German teaching and experience) increased the possibility of applying rational, effective and positive solutions. On the other hand, by the nineteenth century, the central Spanish government had accumulated sufficient legal, administrative and financial power to impose a centralised forest administration that could begin to influence the development of silviculture effectively, although imperfectly.

As a result, an administrative, legal and educational infrastructure was established. The 1833 Forest Ordinances can be considered to mark the starting point of Spanish forest policy, clarifying the responsibility of the government in different woodland areas according to property rights, and prompting the creation of the State Forest Authority in 1837. In subsequent years, the Forest Service was subdivided into planning, district, plantation and hydrological units.[2] The creation of the Forest Engineers Corps in 1853 was a vitally important step in this process, being composed of graduates from the new Forest School opened in 1848.

The new forest administration had to dedicate its first major political and financial efforts to combating policies of privatising common land, which the government was promoting in response to liberal economic ideas and/or to the need to gain liquid cash to pay off debts (Jovellanos 1795; Olazabal 1883). Such privatisation implied the loss of many of Spain's remaining wooded areas, since these were often found in common lands. Beech, oak and pine woods were eventually excluded from the catalogue of lands to be privatised. However, these problems and an insufficient budget meant that very little afforestation was undertaken until the twentieth century, and it was only after the Spanish Civil War (1939) that huge afforestation schemes began to be implemented.

## The characteristics of afforestation in Spain

Analyses of the characteristics of Spanish forest plantations (above all, species composition) indicate that most planting activity should be classified as 'afforestation', and very little as 'reforestation' in the sense of restoring quasi-natural forests. Lively debates amongst foresters underline the fact that afforestation itself was (and is) also open to many interpretations concerning the type of property (state, common land or private) in which it should be undertaken, the types of incentive to be used to encourage the participation of private landowners (grants, loans, subsidized machinery or seedlings, etc.), the type of species and silvicultural techniques to be used in plantation projects, the rhythm that afforestation should try to attain, the priority to be given to the objectives of afforestation (physical, social, cultural, economic-industrial), and the type and degree of integration between plantations and other rural activities (integrated multifunctionality or uni-functional segregation).

Over 3.5 million ha were planted between 1940 and 1983 (Figure 6.1: data vary from one statistical source to another and reflect the sub-optimal state of

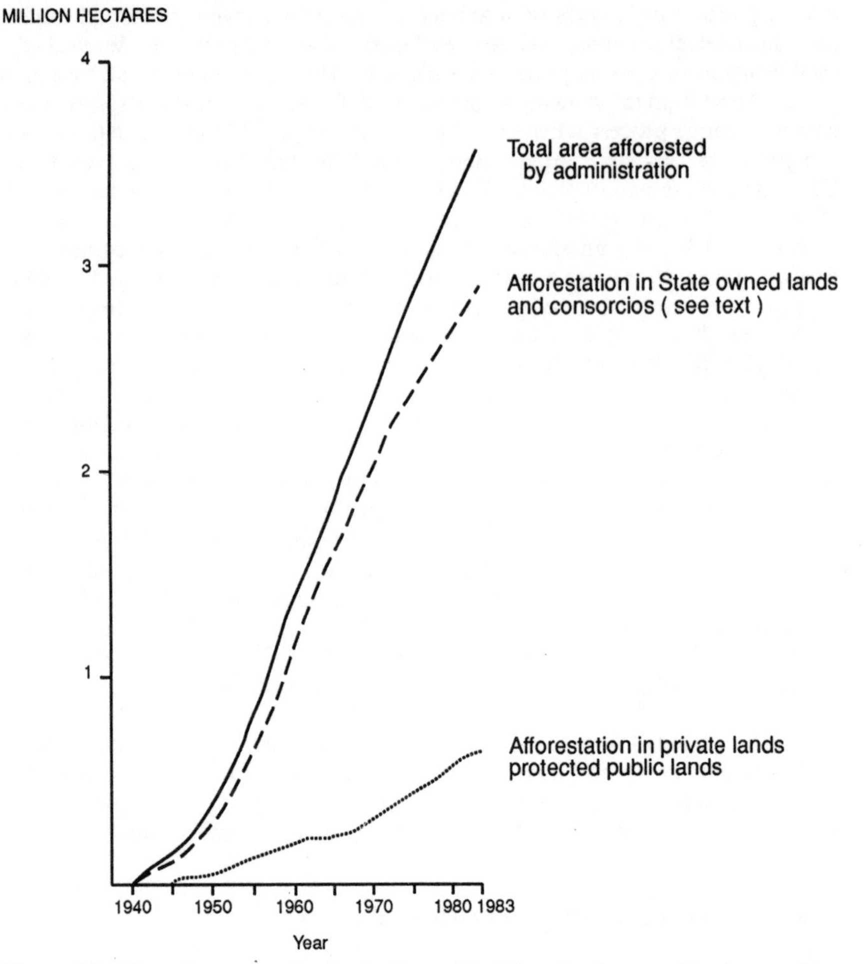


**Figure 6.1** Plantations undertaken by the Forest Administration, by type of land ownership.
*Source*: PFE and ICOA annual reports.

most forest statistics in Spain), and a further 200,000 ha (approximately) since decentralisation of forest activities.[3] The characteristics of these plantations require careful analysis in order to assess afforestation policies in relation to different elements of the economy and environment, above all in rural areas. Afforestation policy can thus be evaluated in terms of its advantages and disadvantages, or the costs and benefits it has generated.

## Woodland ownership and afforestation

Policies developed in the nineteenth century to privatise or 'enclose' common lands absorbed most of the new forest administration's initial activity. Many early foresters, whilst in favour of private property in general, opposed privatisation in their sector. The debates thus generated analysed and evaluated in depth the characteristics of each type of ownership of forest land.

Several schools of thought emerged and the consequences of each for afforestation policy were immediately apparent. On the one hand, those in favour of promoting private woodland ownership argued that free market forces would regulate economic activities in the forest sector, and guarantee afforestation and timber production according to supply and demand on the one hand and the search for maximum revenue on the other (Jovellanos, quoted in Martínez Hermosilla 1942). Optimum resource distribution would increasingly be guaranteed by the market economic criteria that gradually penetrated Spain's rural areas from the nineteenth century onwards, and afforestation policy in particular would thus be guided by timber prices and sales. In strict terms this analysis left little room for incentives, given the contradictions these would create in terms of distorting free market forces.

A second analysis suggested that public and private forestry should be combined, given the particular services to be offered by each: on the one hand general services such as the control and prevention of soil erosion, flood prevention and salubrity that appeared in the texts of early foresters (Fivaller 1868; García Martino 1868; Alvarez Sereix 1884), and in more recent terms the maintenance of species and genetic diversity; and, on the other hand, special services which included the production of timber, firewood, charcoal and woodland pastures. Clearly, in modern terminology, the division refers to the generation of positive externalities on the one hand and market values on the other.

According to this school of thought, given that private woodland owners are interested in obtaining maximum income, they rarely maintain or plant woodland areas in which the principal function of tree cover is, for example, to prevent erosion. For this reason, those areas in which woodlands exist or should exist to guarantee general services of one sort or another should be publicly owned. Such areas were initially 'protection' woodlands and afforestation (and maintenance) and these would be in the hands of the forest authorities (Senador 1933).

On the other hand, it was thought that certain woodlands only needed to provide special services: in other words, to produce timber and other forest products. In these areas, market forces and the right to private property were to be respected and they were termed 'production' woodlands. Afforestation and woodland management would therefore be in the hands of private landowners.

The peculiarities of forestry economics influenced the determination of a third school of thought. Private initiative was considered to be in contradiction with the long production cycle that characterises most tree species[4] (Jordana 1870; Pascual 1870), and therefore even woodlands offering only special services or fulfilling production functions should be socialised in one way or another. Additionally, it was considered that private woodland owners could not guarantee a regular, sustained timber supply. Felling would tend to be adjusted to market prices, with high prices causing a tendency to increased felling and vice versa (Garcia Martino 1871; Griñan 1917), without ensuring either regular industrial supply or subsequent regeneration of wood supplies whether in afforestation or reforestation schemes.

Lastly, certain groups opposed private property and inheritance rights in principle (and not only in the forest sector). In this case, private woodland ownership was not thought capable of providing either 'general' or 'special' services (Fivaller 1868). The alternative offered to private woodland ownership was not, however, simple nationalisation. Throughout the tumultuous years of Spanish restoration, revolution, dictatorship and republicanism (1875–1936), some tendencies favoured state ownership whilst others favoured the maintenance and promotion of the variety of communal forms of land ownership that still persist in Spain today. Accordingly, woodland management and afforestation could be a state-controlled activity or a more decentralised 'public' activity.

Yet others proposed limited state intervention in private woodlands, in order to orientate management of the 'capital' forest vegetation represented, whilst maintaining private ownership of land (Reyes 1917). As Fenech (1917) insisted, land ownership rights were not to be withdrawn but rather restricted by certain limitations, as a logical consequence of the relationships that exist between the individual, the state and society. Woodland management and afforestation would thus be a state-supported activity in financial and infrastructural terms.

In subsequent decades the reality of afforestation practice confirmed the contradictions existing between private woodland ownership and the maintenance or promotion of slow-growing tree species. It did not necessarily ratify the theories concerning public ownership and the provision of 'general services', since industrial and financial criteria also tended to dominate the orientation of public afforestation schemes, to the detriment, in fact, of both these 'general' and certain 'special' services.

Plantation programmes were thus predominantly state-promoted, but did not imply the freezing of capital in land acquisition (or nationalisation with compensation), but rather state management or control of the capital represented by forest vegetation, employing two major mechanisms.

## Mechanisms of afforestation

### *Consorcios*

First, these programmes afforested common lands following the signing of agreements (known as *consorcios*) with village councils. As can be seen in Figure 6.2a, such agreements provided the bulk of the lands to be afforested during the last four or five decades in Spain (compare with Figure 6.1). However, the imposition of such agreements during the Franco regime generated a whole series of conflicts that have had serious consequences for the success of the plantations. No provision was made to create alternative grazing lands for livestock farmers, who relied heavily on common-land resources for forage. And only 270,000 ha were subject to activities of pastureland creation, maintenance and restoration, during a period in which 3.5 million ha were afforested (Figure 6.2b).

The financial implications and opportunity costs of such plantations were questioned by foresters (and members of other professions) who, despite being in favour of plantation programmes, stressed the need for the integration of forestry with livestock grazing in particular and the rural economy in general (plantations were said, by some, to generate more and, by others, less employment than

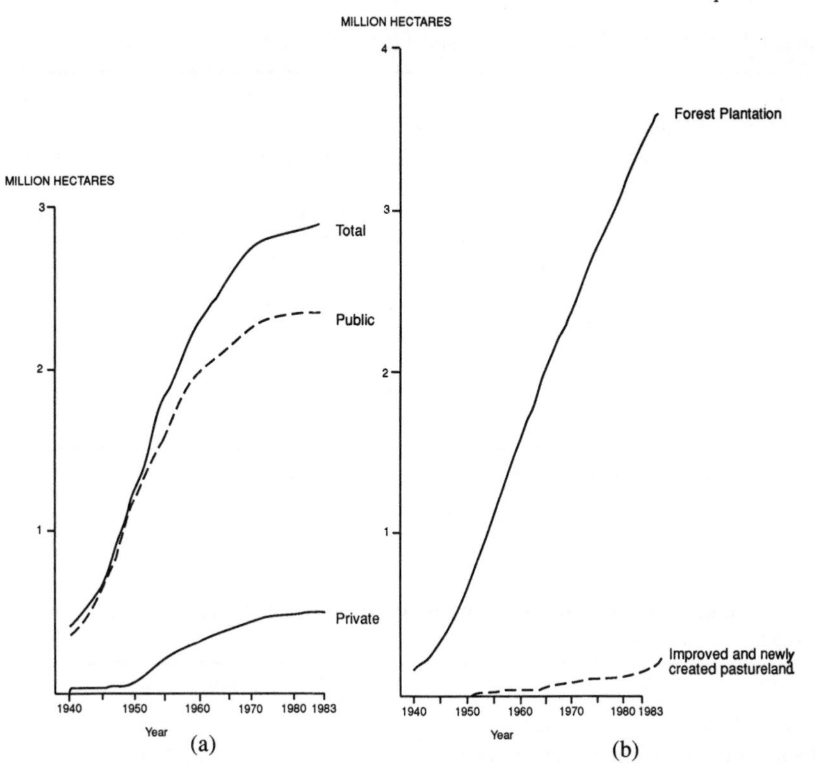


**Figure 6.2a** Cumulative area of '*consorcios*' by type of land ownership.
**6.2b** Cumulative area of pasturelands created or improved 1945–83, compared with forest plantations.
*Source*: ICONA annual reports.

traditional multiple-resource exploitation of common lands (Bernard 1963; García and Marquez 1978), whilst the loss of grazing lands meant more financial resources had to be spent on imports of forage (Tejerina 1958)).

Heavy fines penalised those shepherds and goatherds trying to recover traditional grazing rights. Simultaneously, rights to gather firewood or material for producing charcoal, and all secondary products such as mast, honey, and mushrooms, were lost, despite the fact that they constituted important support products in the local rural economy. Experience has at times proven the vital importance of secondary products in forestry. For example, prohibition of woodland grazing meant dead biomass rapidly accumulated in new plantations, increasing the propensity and ease with which forest fires were propagated.

Resistance to the loss of communal resources was a major factor in the low success rate of many plantations, although numerous other factors such as inadaptation of species have also played their part. The setting of forest fires (in order to reclaim pasturelands) was predicted by some foresters who warned of the social consequences of the mass afforestation underway in common lands (Laffitte 1919). The *consorcios* could, in fact, be considered to be a form of

**Table 6.1** Elements of private forest capital to benefit from incentive policies

|  |  |  | |
|---|---|---|---|
| CIRCULATING CAPITAL | | *Seeds, plants, insecticides, technical administration, insurance, taxes,* fertilisers, security, labour force | |
| FIXED CAPITAL | TERRITORIAL | Forest soil, vegetation<br>Rural buildings, walls, boundaries<br>*Forest rides, firebreaks, terraces,* ditches | |
| | EQUIPMENT | Mechanical | Hand tools<br>*Heavy machinery* |
| | | Live | Animals |

* Elements traditionally covered by incentive programmes
*Italics* indicate availability of financial support.
*Source*: Based on Elorietta and Artaza (1920) *Principios de Economía Forestal Española*, Tome 1, p. 160.

Relative values of incentives have changed over time. In the province of Vizcaya, levels of support in 1991 were:

| Activity | Percentage level of support |
|---|---|
| Planting slow-growing broadleaves | 85 |
| Planting slow-medium growing decid./conifer | 60 |
| Planting fast growing spp. (*Eucalyptus, P. radiata, P. pinaster, Populus*) | 30 |
| Enclosure of plantation | 45 |
| Thinning, clearing undergrowth | 40 |

resource privatisation, a view encouraged by the narrow range of species and industries favoured in afforestation.

### *Private-sector incentives*

The other major mechanism used by the forest administration to encourage planting was a package of incentives for woodland owners to promote afforestation in the approximately two-thirds of Spain's forest area that the private sector owned (ICONA and DGPA 1987). (Perhaps surprisingly, there are still no definite figures regarding the exact percentages of different types of forest property ownership in Spain.) These incentives covered most forms of forest, as can be seen in Table 6.1, but, as will be shown, tended to encourage the use of quick-growing tree species. Adequate provision was not made to overcome the long period between investment and amortisation that characterises most native Spanish species, which were also marginalised in industrial policy.

The state did eventually purchase about 1 million ha of forest land between

1940 and the 1980s. However, this was not only for afforestation purposes but also to guarantee the proper management of national parks, and the financial outlay in land acquisition represented a small amount in comparison with investment in afforestation (4,700 and 33,407 million pesetas, respectively: ICONA, various years; Naredo and Marquez 1987).

## Species composition of the plantation programme

Various factors have influenced the species composition of the afforestation undertaken in Spain. One is the contradictions between new market economic criteria and the growth cycles of many tree species, with continual pressure for shorter investment cycles. From the nineteenth century exotic species (especially *Pinus radiata* and *Eucalyptus* spp.) were widely used, with much shorter growth cycles than most native tree species. Certain species were deliberately promoted at the expense of others in packages of grants and subsidies. And lastly, the choice of species was affected by the rapid development of the cellulose industries, above all after the Spanish Civil War (1940s onwards). These industries were positively encouraged and favoured by the Forest Administration.

Whilst certain members of the Forest Engineering Corps favoured the almost complete predominance of a relatively small number of species in plantations, to take advantage of quick growth cycles and/or to guarantee supplies to the growing, state-aided paper industry (Adan de Yarza 1913; Lillo 1931; Echevarría Ballarín 1932; Martínez Hermosilla 1947; Rojas 1972), others warned of the various economic, social and environmental repercussions such a policy could and did provoke (Villacampa 1868; Ventallo 1908; Laffitte 1919; Robles Trueba 1932; Kozdon 1954; García Diaz 1957; Ramos Figueras 1958).

It was the private sector that had commenced experiments with quick-growing species (such as *Eucalyptus* spp.) in the nineteenth century in Spain (Echevarría Ballarín 1933a). Many foresters praised their short-growth cycles (Aristegui 1868; Caro 1920), and the Department of Agriculture soon organised the distribution of eucalypt seeds to both private and public landowners. Initially it was thus thought that the contradiction between private woodland ownership and the maintenance of adequate timber supplies ('special services') had been overcome (Echevarría Ballarín 1933b).

Most private landowners opted for quick-growing species, given the need to minimise investment cycles in order to maximise financial returns. Interest rates and opportunity costs weigh heavily in the forest sector since, even today, the shortest production (or investment cycles) are over eight years long (the mini-rotations that characterise some eucalypt monocultures) and tend to be 15–20 years or 25–40 years (for most *Eucalyptus* and nearly all quick-growing *Pinus* species, respectively). Plantations with slow-growing species under current economic conditions and with existing incentive schemes mean that returns on investment are generally low, and are only available in one or two generations' time.

Additionally, the lack of income during the production cycle, apart from possible revenue to be derived from thinning, means that small landowners must combine forestry with other activities (industrial employment or other agrarian

**Table 6.2** Species composition of plantations undertaken by the private sector with official support, 1968-87

| Year | Area (ha) | Eucalypts | Poplars | Slow-growing deciduous | Conifers* |
|---|---|---|---|---|---|
| | | | (per cent of area) | | |
| 1968 | 868 | | | | |
| 1969 | 4,042 | | | | |
| 1970 | 4,030 | | | | |
| 1971 | 11,151 | 87.4 | 4.0 | 0.6 | 8.0 |
| 1972 | 17,278 | 91.3 | 2.7 | 0.1 | 5.9 |
| 1973 | 15,441 | 84.3 | 5.6 | 0.8 | 9.3 |
| 1974 | 14,837 | 77.0 | 9.3 | 0.7 | 13.0 |
| 1975 | 21,927 | 69.8 | 10.4 | 1.9 | 17.9 |
| 1976 | 29,828 | 61.4 | 9.4 | 1.2 | 28.0 |
| 1977 | 40,030 | 70.8 | 7.2 | 2.5 | 19.5 |
| 1978 | 21,355 | 51.7 | 13.4 | 5.4 | 29.5 |
| 1979 | 22,566 | 49.4 | 12.2 | 4.3 | 34.1 |
| 1980 | 12,452 | 32.3 | 20.6 | 6.3 | 40.8 |
| 1981 | 17,674 | 30.2 | 21.5 | 10.8 | 37.5 |
| 1982 | 12,901 | 40.7 | 22.5 | 6.1 | 30.7 |
| 1983 | 11,672 | 19.2 | 22.9 | 18.9 | 39.0 |
| 1984 | 7,842 | 13.6 | 28.1 | 17.9 | 40.4 |
| 1985 | 4,477 | 19.4 | 23.3 | 13.2 | 44.1 |
| 1986 | 11,758 | 26.2 | 23.0 | 19.8 | 31.0 |
| 1987 | 10,872 | 24.8 | 18.0 | 26.7 | 30.5 |

* Includes some aromatic plants (e.g. 1976 = 0.7 % of total)
*Source*: Ministry of Agriculture, *Anuario de Estadística Agraria*

activities) in order to receive a monthly or yearly income whilst waiting for income from forestry. They therefore generally plant quick-growing species in order to supplement their income within their own life-cycle.

As a result most private plantation programmes constituted 'afforestation' as opposed to 'reforestation' schemes, as the species used tended not to replace former, slow-growing vegetation but were new. The forest administration positively encouraged this trend from the nineteenth century onwards, making the seed and plants of such species readily available to private owners and focusing all research into reproduction, growth cycles and disease on quick-growing species (Elorrieta 1960; Michel 1986).

The type of species used in plantations supported by official aid is shown in Table 6.2 (slow-growing species constitute more than 10 per cent of private-sector planting sector only from the 1980s, and exceed 20 per cent only once).

The financial benefits to be gained from quick-growing species also encouraged afforestation by the private sector even without financial support. In the Basque country (northern Spain), for example, small landowners and others have planted over 100,000 ha under *Pinus radiata* over the last four or five decades, with 25–40 year production cycles.

**Table 6.3** Species composition of plantations undertaken by the Forest Administration, 1940–83 and 1984–87*

| | 1940 – 1983 | | 1984 – 1987 | |
|---|---|---|---|---|
| | Ha | % total | Ha | % total |
| CONIFERS | | | | |
| *P. pinaster* | 794,478 | 26.3 | 29,005 | 16.6 |
| *P. sylvestris* | 565,489 | 18.7 | 30,481 | 17.4 |
| *P. halepensis* | 485,814 | 16.1 | 43,298 | 24.8 |
| *P. laricio* | 385,986 | 12.7 | 12,191 | 7.0 |
| *P. pinea* | 222,826 | 7.4 | 28,891 | 16.5 |
| *P. radiata* | 168,909 | 5.6 | 8,891 | 5.1 |
| *P. canariensis* | 27,453 | 0.9 | 1,567 | 0.9 |
| *P. uncinata* | 17,807 | 0.6 | ? | |
| Other conifers | 22,944 | 0.8 | ? | |
| *Populus* spp. | 26,544 | 0.9 | 3,955 | 2.3 |
| *Eucalyptus* spp. | 273,877 (532,000)† | 9.1 | 767? | 0.4? |
| All other broadleaves | 25,788 | 0.9 | ? | |

*Control of forestry had been transferred to most regional governments by 1984.
† The Pulp and Paper Industrial Association estimates total eucalypt plantations to reach 500,000 ha by the 1980s – including many plantations on private land.
*Sources*: ICONA *Annual Reports*; Ministry of Agriculture, *Annuario de Estadística Agraria*

No positive incentives were developed in Spain to promote slow-growing species in private woodlands. Even the offer of 100 per cent subsidies for plantations has failed, because no income is gained during the investment cycle. The possibility of making the financial benefits of these plantations available to landowners during the growth cycle by offering an annual rent for those landowners planting slow-growing species is being debated both in the EC and by rural agents in Spain (Commission of the European Communities 1991; Confederación EHNE-UGAV 1991) as the only viable mechanism, either as an alternative to traditional farming or as a complementary activity.

However, it should be stressed that the financial returns of even quick-growing species are deteriorating as falling timber prices, high interest rates and growing input costs whittle away margins. The unit costs of planting rose from 2267 ptas/ha in 1958 to 4177 ptas in 1977 (real costs), whilst the average real (constant) price of wood fell from 742 ptas/m$^3$ in 1958 to 334 ptas in 1970 (Mateo Sagasta 1979). Recent data confirm these trends (Naredo and Marquez 1987; Gobierno Vasco 1991). Thus the idea that the contradiction between private forestry and the maintenance of tree cover had been overcome by the use of quick-growing species has lost ground in the face of high interest rates and low wood prices. Private-sector afforestation with quick-growing species therefore depends more than ever on official subsidies or new silvicultural techniques that attempt to cut planting costs, the negative externalities generated by the latter being covered by society.

Despite the different characteristics of public and private forests and the 'general' and 'special' services these should supply, the species composition of both private and public plantation schemes has been remarkably similar. From the start of the nineteenth century, forest legislation and practice increasingly prompted quick-growing species even on public lands and even in those areas in which the maintenance or regeneration of natural vegetation would adequately fulfil protection functions.

The outcome has been the complete marginalisation of typical forest species such as oaks (Mediterranean and Atlantic species), beech and chestnut (a species thought to have been introduced by the Romans) precisely because of their slow growth cycle (Table 6.3). In fact less than 1 per cent of all planting undertaken by the government itself employed slow-growing broadleaved species, and over 40 per cent employed quick-growing species. Conifers dominated planting programmes, despite their secondary importance in Spanish forest vegetation. Some 90 per cent of all plantations employed pines, some of slow but most of medium and quick-growing species, mainly for supply of raw materials for the pulp and paper industries. The biggest concentrations of such plantations are found in the northern Atlantic areas of Spain, although eucalypts were also planted in Andalucia and Extremadura and pines of one sort or another throughout the country.

The type of plant and seed available in official nurseries reflected this pattern: between 1940 and 1984, 97 per cent of both plant and seed available were of conifer species. Of the angiosperm plant available, most were of eucalypt and poplar species. As the Forest Administration itself admitted, only a few small plots of other broadleaved species were planted 'but rarely for integration in afforestation projects' (Giménez Radix 1950). In the meantime, vital knowledge concerning the reproduction and growth cycles of slow-growing species and traditional regeneration techniques was gradually lost and no research was undertaken to further knowledge.

The very poor success rate of the Spanish afforestation programme suggests that early calls for adequate experimentation with exotic species before their mass implementation and the need to guarantee a variety of species within each spatial unit of afforestation in order to reduce the risk of disease should have been heeded. It has been estimated that between 0.85 and 1.2 million ha of the plantations never became properly established (Naredo and Marquez 1987), with obvious financial and environmental implications. The use of biogeographically unsuitable species or varieties was a major factor. Additionally, an estimate of the total wooded area lost due to forest fires, disease, pests and wood mining (exploitation without replacement), reveals that, despite the afforestation programmes undertaken, Spain's wooded area has actually diminished by about 1 million ha since 1947 (García Dory *et al.* 1984).

It is thus clear that afforestation policies should not be judged by the number of hectares planted per year, as is the usual case in both national and international literature, but rather by their long-term success rate, which depends on, amongst other parameters, the balanced mix of species employed, the techniques used in plantations, the multiple interests they satisfy and the positive externalities they generate.

## Silvicultural techniques used in afforestation

Traditional 'natural' silviculture has gradually given way to 'artificial' or 'orchard' silviculture (Kalela 1960; Leibundgut 1960) worldwide, not least due to the type of mechanical and chemical techniques and products incorporated into forestry with the need to reduce labour costs and increase productivity. However, the rate and degree to which such techniques and products are incorporated into a new sector may exceed advisable limits, not only in terms of the physical capacity of given lands to absorb the impacts they produce, but also because of their social consequences and even long-term economic consequences.

The Spanish Forest Administration commenced mechanisation of forest activities in the 1950s, stressing the advantages to be gained from a reduction in the labour force (at a time when official propaganda also claimed that plantations were being promoted to provide rural employment); specialisation of the labour force; an increase in the speed at which afforestation could take place; a reduction in the number of badly planted trees; a reduction in costs of inspection; and a reduction in overall plantation costs (Abreu 1965; Nájera 1962; Ruiz-Tapiador 1965). The Forest Administration itself was authorized to purchase forest machinery (mainly from central or northern Europe initially), and made grants available for its purchase by the private sector.

One of the main problems provoked by this process was the lower level of rural employment thus generated: plantations in previously grazed areas have usually provided less employment than the former land use. Another issue arising from mechanisation was a range of environmentally-related problems with economic undertones. The indiscriminate use of machinery in all phases of plantation management (land preparation with heavy machinery, including the downhill subsoiling of steep slopes; clear cutting of large areas, including, once again, slopes; the use of heavy machinery to log and haul wood from plantations; and the sub-optimal development and very poor upkeep of a dense network of forest tracks) meant both the negative externalities and complaints generated by forest activities multiplied dramatically. Scientific studies confirm the environmental implications of such inadequately employed techniques (e.g., erosion, soil loss, slope destabilisation, water quality loss (Pablo *et al.* 1991)).

The increasing use of chemical products in the form of both insecticides and fertilisers (only the latter are now being used on a large scale) has also provoked reaction. Previously products such as DDT were employed on a spatially widespread basis until caution prevailed at an international level. New chemical insecticides are now in use (Diflubenzuron, for example) as calls for effective prevention (through an end to monocultures, for example) rather than the elimination of forest pests have still to be adopted. The philosophical and scientific implications of different viewpoints (short-term versus long-term side effects or specific or widespread action, for example) still fuel debate.

## Afforestation and supply of the wood industries

The different qualities that characterise the wood produced by each class of trees – physico-mechanical resistance, colour, texture, durability, fibre length – were recognised by early foresters (Olazabal 1856), who commented, for example with regards to eucalypts, that 'the vertigo with which wood is produced means quality is sacrificed' (Lleó Silvestre 1929). Aracil Laborda (1970) concluded that pine, the most abundant wood available in Spain following the mass plantation programmes, was generally of low or medium quality. The exception, wood from certain areas of Scots pine, corresponds to early plantations established before the mass projects of the Franco era. The implications for the forest manufacturing industry have been clear.

Close feedback relationships between afforestation and forest-industry policies meant the development of the two major sectors of the timber-consuming businesses in Spain was completely unbalanced. Those industries relying on high-quality wood for use in its solid form for furniture, carpentry and building were gradually forced to search for resources abroad or use the lower-quality woods produced in Spain, not only because of the species composition of afforestation programmes but also because of a simultaneous abandonment of remaining natural wood units in most of Spain. Simultaneously, the lack of adequate protective and regeneration policies meant that these remaining natural woodlands continued to be exploited and depleted. Native hardwoods were eventually complemented or substituted with species such as American oak, European beech and a range of tropical species, but not without complaints by this sector of the wood industry and calls for the use of a wider variety of species in plantation programmes (Muñoz Goyannes 1958; Piera 1985).

On the other hand, those industries that disintegrate wood and then reconstitute its fibres in a variety of products (basically the pulp, paper and artificial board industries) were promoted with different forms of aid (high export subsidies, special grants for factories, research: see Groome 1990 for a review of the relevant legislation), apart from being able to make use of the vast majority of species used in afforestation programmes.

As can be seen in Table 6.4, the percentage of wood produced in Spain and supplying the fibre industries has gradually increased during the last three decades, whilst the solid-wood industries have lost their traditional importance. The recent relative increase in supplies to the solid-wood industry is the result of a campaign by pine growers to expand their markets and to supply not only the paper industries. However, wood quality must be considerably improved, according to industrialists from the solid-wood sector (Groome 1991) in order to make the use of timber from today's plantations generally acceptable in high value-adding industries (genetic improvement of seeds and plants, careful silvicultural management and post-felling wood treatment).

Despite the vast amount of wood supplying pulp and paper industry in Spain, over 46,610 million pesetas (£233 million) are spent each year in imports of raw materials for this industry, since the plantations have not been able to produce enough wood for the huge production capacity installed. The feedback mechanisms at work are now, therefore, promoting the planting of even more lands with *Eucalyptus*, *Populus* and *Pinus* species, in order to meet demand. Simultaneously, however, Spain also exports pulp, mainly to EC countries, since production capacity is not orientated to domestic needs. Exports are of about the same total value as imports (46,255 million pesetas). The export of pulp (and not paper) means value added is not maximised, whilst the most polluting wood

**Table 6.4** Consumption of domestically produced wood in Spanish wood manufacturing industries, 1955–88

| Industrial sector | % total classified wood produced | | | | | | |
|---|---|---|---|---|---|---|---|
| | 1955 | 1963–6 | 1972 | 1978 | 1984 | 1987 | 1988 |
| Pulp for paper | 7.2 | 25.7 | 32.6 | 32.0 | 40.6 | 46.3 | 37.0 |
| Artificial boards | – | 8.7 | 9.6 | 11.0 | 10.8 | 9.7 | 11.0 |
| TOTAL DISINTEGRATED | 7.2 | 34.4 | 42.4 | 43.0 | 51.4 | 56.0 | 48.0 |
| Veneer | – | 1.6 | 0.7 | – | 4.4 | 3.1 | 4.4 |
| Sawnwood* | 55.4 | 37.0 | 43.6 | 43.0 | 36.7 | 35.6 | 42.4 |
| Other** | 37.4 | 27.0 | 13.6 | 14.0 | 7.5 | 5.2 | 5.2 |
| TOTAL SOLID WOOD | 92.8 | 65.5 | 57.9 | 57.0 | 48.6 | 44.0 | 52.0 |

* Including packaging
** Posts, sleepers, etc.
*Sources:* Compiled from various sources, including: Ministry of Agriculture, *Anuario de Estadistica Agraria*

industry is being promoted. In the meantime, no definite supply-based policy exists for the other sector of the forest industry (which is also characterised by raw-material deficits in the EC), other than the aforementioned attempts to transfer some pinewood supplies to the solid-wood industries.

Little attention has thus been paid to the pleas of earlier foresters to maintain stocks of all tree species not only for physical or environmental reasons but also to respond adequately to changing demands (González Vázquez 1938). Those industries needing mechanico-structurally resistant woods, large-sized beams, or a wide variety of decorative textures and colours, for example, now face difficulties of supply, with no prospect for immediate solution given that the species supplying these woods are still not being planted to an adequate degree.

## Consequences of regional devolution for forestry

Regional devolution of political powers in general took place during the 1980–1985 period in Spain. Decision-making in forestry, as part of agrarian policy in general, was also transferred to the regional governments set up throughout Spain. In general, until the 1990s no practical changes were observed in plantation policy, as, despite new legislation improving the financial incentives for planting slow-growing species, these were still unable to overcome the long period that persists between investment and amortisation and that thwarts private initiative. Additionally, many plantations in public lands continued to employ conifers and/or quick-growing species.

During recent years, some changes have occurred, but in the opinion of some Environmental Departments and environmentalist groups, slow-growing species are still marginalised when compared with the research undertaken into quick-growing species, with the different availability of plants in nurseries (measured in terms of price, secure supplies, quality, etc.) and with the area and quality of plantations undertaken (adequate, up-to-date official figures are not available but plantations with slow-growing species may now amount to several tens of thousands of hectares).

The silvicultural techniques being used in forestry since the transfer of decision making to regional governments have changed, but generally in terms of a greater emphasis on mechanisation, and there has been no significant change in the use of chemical pesticides. Equally, there has been little change in the incentives given to promote slow-growing species, and as mentioned, the few changes produced have not had results at a practical scale.

The most encouraging development is the tendency to incorporate the views of different social groups in the formulation of 'Forest Plans' in some new regional governments (Andalucia, Galicia, Navarre), a process that should be encouraged to reduce the number of conflicts caused by planting and to maximise benefits. These plans set out a proposed zoning, aimed at protecting existing remnants of natural woodland and designating areas to be afforested or reforested, and with what species.

## Perspectives for afforestation and reforestation in Spain

A growing number of the many groups that constitute Spanish society are calling for urgent changes in planting programmes. If carried out, such changes would effectively convert many afforestation programmes into reforestation schemes. The basic calls are related to three major features, outlined above, that characterise the afforestation schemes that have marked Spanish forest policy to date. In the first place, there are calls for the use of a greater variety of species, including, above all, native slow-growing species, not only from conservationist groups but also from farmers' unions, consumer associations, beekeeper associations, political parties and, of course, those industries which consume good-quality timber (Groome 1991). The species most named by all such groups for use in a new programme of plantations are slow- (or medium) growing broadleaved species (beech, walnut, cherry, ash, oak, etc.). The fact that these species are also contemplated in plantation scenarios designed by the Club de Bruxelles reveals the interest of solid-wood using timber industries further afield (Smith 1990).

It is important to note that the same research revealed a certain number of industrialists who recognised that such plantations would not only solve the long-term supply problems facing their industries (due to the lack to a domestic supply, due to the 'mining' of tropical forests, export limitations and the promotion of domestic transformation industries in producer countries, etc.) but also mentioned environmental reasons for their promotion. On the other hand, most groups such as environmental associations not only stress the need for species diversification in order to recover species diversity and related habitats and landscape values, but also to ensure industrial supply.

Secondly, calls are made by industrialists for changes in silvicultural practices in order to guarantee the optimum quality of wood produced in Spain. Pruning, thinning and periodical clearance of forest undergrowth maximise wood quality and help reduce the risk of pests and fires, a fact increasingly stressed by other groups such as environmentalists, farmers' unions and researchers. Such management practices have been traditionally neglected in both private and public plantations, due to both a lack of profitability and the lack and loss of relevant know-how amongst private foresters. More careful timber extraction would also reduce the loss of timber value during logging.

In general, calls are also being made for changes in plantation techniques, in order to reduce the enormous environmental costs they currently produce. The use of downslope subsoiling and clearcutting on steep slopes are two of the most criticised techniques currently used by the Forest Administration in plantation programmes. They not only cause erosion and loss of fertility in the areas in which they are practised but also lead to sedimentation of streams, reservoirs and ponds, loss of water quality and the disruption of hydrological cycles.

Thirdly, the different social groups contacted in recent research are in favour of overcoming the apparent contradiction between private land ownership and plantation programmes with species of all growth rates. Land nationalisation is not acceptable, but a new form of 'socialisation' is contemplated through a mechanism in which landowners are encouraged to plant slow-growing species – the combined use of subsidies to cover plantation costs and the advancing of the income these would provide in a hundred or more years time in the form of an annual rent – but which imposes on the landowners the obligation to manage forest land appropriately, employing the species, silvicultural techniques and management plans best suited to the particular physical conditions of each area in order to generate positive and minimise negative externalities.

The concept is surprisingly similar to the view already quoted that private foresters be respected but regulated 'as a logical consequence of the relationships that exist between the individual, the state and society' (Fenech 1917).

## Conclusions

Plantation policies in the past in Spain have clearly been orientated towards afforestation, reforestation in the sense of restoring quasi-natural forests playing a very marginal role in silviculture until very recently.

Afforestation has been a mainly state-promoted activity, with private initiatives relying increasingly on incentives of one sort or another in order to plant, given the economic restraints that limit the economic viability of most forest-production cycles. Both private and public afforestation schemes have promoted quick-growing and/or conifer species, whilst slow-growing, mainly broadleaved species have been completely neglected despite constituting the typical vegetation of both Atlantic and Mediterranean Spain, and despite the existence of physical, social and industrial reasons for their adequate integration in plantation programmes.

The silvicultural techniques employed in forestry have caused a growing amount of scientific and social concern, in the light of the high number of negative externalities generated and the loss of numerous positive externalities. Afforestation policy and practice favoured supply to wood-disintegrating industries, and especially to the cellulose industry. Solid-wood using industries, such as furniture, were forced gradually to use sub-optimal quality woods or to search for timber abroad.

Industrial, rural and social interest groups have joined in calling for change in species composition, financial incentives and silvicultural techniques employed in forestry, which would open the door to reforestation schemes without renouncing afforestation.

Optimum mechanisms for the minimisation of conflicts and maximisation of benefits in the forest sector, such as the preparation of forest plans at different geo-administrative scales, integrated into general rural-planning frameworks, and which incorporate rather than marginalise the many interest groups that exist in today's Spanish society, should be positively encouraged.

## Notes

1 The Mesta, an organisation which grouped transhumant shepherds and sheep owners, obtained tremendous powers during the Middle Ages. It fought for rights to pasture and grazing, drove roads and other aspects of transhumant infrastructure. Some historians claim that the Mesta was responsible for overgrazing, deforestation and subsequent land degradation. Other historians claim that the Mesta was fundamental in that it guaranteed the existence and maintenance of drove roads, a vast network of which still exists in Spain and which enabled complementary use of winter and summer grazing lands, thus helping to maintain a livestock economy that would have been unable to survive with only summer or winter lands. This, in turn, helped to enhance land and ecological diversity (see Klein 1920 for further details).

2 The forest administration has since undergone successive reforms:
1935–41 – introduction of the State Forest Heritage (Patrimonio Forestal del Estado (PFE)), dedicated almost exclusively to the development of huge afforestation schemes.
1971 – PFE replaced by the new 'nature conservation' organisation ICONA, although continuity was stressed in afforestation policy.
1980–85 – decentralisation of forest policy and decision making to regional 'autonomous' governments throughout Spain.
1986 – Spain joins EC, the forest sector becoming subject directly or indirectly to EC regulations.

3 These areas (3.5 + 0.2 million ha) compare with approximately 7 million ha of high forest and 16–20 million ha of total forest and woodland, including scrub.

4 The most representative tree species in both Mediterranean and Atlantic Spain are slow-growing species of oak and pine, with growth cycles of around 100 years.

## References

Abreu, J. (1965) Meditaciones sobre la silvicultura y la ordenación de los montes de utilidad pública, *Montes* 21 (123): 217–22.

Adan de Yarza, M. (1913) 'La repoblación forestal en el País Vasco'. Reprinted in *Euskadi forestal*, 1984. Asociación de Propietarios de Forestalistas del País Vasco, Bilbao, 24 pp.

Alvarez Sereix, R (1884) *Estudios botánico-forestales*. Ginés Hernández, Madrid, 103 pp.

Aracil Laborda, A. (1970) La utilización industrial de los recursos forestales del país, *Economía Industrial* 77: 9–17.

Aristequi, E. (1868) Crónica: *Eucalyptus globulus, Revista Forestal* 1: 61–3.

Bauer Mandershied, E. (1980) *Los montes de España en la historia*. Ministerio de Agricultura, Madrid, 610 pp.

Bernard, A. (1963) Consideraciones sobre el problema de la repoblación forestal de España, *Montes* 19(109): 59–66.

Caro, E. (1920) El eucalipto: sus aplicaciones industriales y su porvenir en la utilización de nuestros baldíos, *Revista de Montes* 44 (1031): 3–11.

Commission of the European Communities (1991) *The development and future of the CAP*, COM(91)258. CEC, Brussels.

Confederación EHNE-UGAV (1991) *Nuevo plan de ayudas para una política forestal diversificada*. Discussion paper, Comisión Forestal de la Confederación EHNE-UGAV. Bilbao, 8 pp.

Echevarría Ballarín, I. (1932) Repoblación forestal aplicada a la industria papelera, *Montes e Industria* 15: 398–402.

Echevarría Ballarín, I. (1933a) El interés privado y la repoblación de los montes, *Montes e Industria* 28: 88–91.

Echevarría Ballarín, I. (1933b) 'Fomento de la repoblación forestal entre los particulares y empresas', in Asoc. de Ing. de Montes, *Aportaciones a la política forestal de España*. Rivadeneyra SA, Madrid, pp. 93–124.

Elorrieta, J. (1960) Labor de mejora de *Populus* realizada en España, *Proceedings V World Forestry Congress* (Seattle), pp. 700–2.

Fenech, J.M. (1917) *Los montes de propiedad particular en los países latinos de Europa*. Editorial Barcelonesa SA, Barcelona.

Fivaller, J.M. (1868) ¿Deben venderse los Montes del Estado?, *Revista Forestal* 1: 234–39.

García Diaz, E. 1957. El problema de las especies nobles asturianas, *Montes* 13(78): 391–93.

García Dory, M.A., Llorca, A. and Pristo, F. (1984) Evolución de la superficie arbolada de España duranta el periodo 1947-1975, *Quercus* 13: 9–14.

García, F.F. and Marquez, D. (1978) Análisis socio-económico de las repoblaciones de eucalipto en Sierra Morena, *Actas Jornadas Trabajo sobre Eucalipto*, November, 55–123.

García Martino, F. (1868) Introducción a la Primera Revista Forestal, *Revista Forestal* 1: 3–16.

García Martino, F. (1871) La destrucción de los montes, *Revista Forestal* 4: 79–82.

Giménez Radix, L. (1950) Labor desarrollada por el patrimonio forestal desde su creación hasta finales del año 1949, *Montes* 6: 367–86.

Gobierno Vasco (1991) *El sector agroalimentario 1990 CAPV*. Vitoria, 159 pp.

González Vázquez, E. (1938) *Fundamentos naturales de la selvicultura*. IFIE, Valencia, 470 pp.

Groome, H.J. (1990) *Evolución de la politica forestal en el estado español*. Agencia de Medio Ambiente de la Comunidad de Madrid, Madrid, 336 pp.

Groome, H.J. (1991) *Repercusiones territoriales en la Cornisa Cantábrica de cambios en el suministro de materias primas forestales a la industria maderera no-celulósica.* Direccción General de Investigación Científica y Técnica. Madrid, Ministerio de Educación y Ciencia.

Griñan, O. (1917) Intervención del Estado en los montes particulares, *Revista de Montes* 41 (981): 828–32.

ICONA (various years) *Los incendios forestales en España.* Ministerio de Agricultura, Madrid.

ICONA and DGPA (1987) Política forestal e incendios forestales, Apéndice I, *Jornadas Política Ambiental*, 142.

Jordana, J. (1870) Desamortización forestal, *Revista Forestal* 3: 256–69.

Jovellanos, G.M. (1795) *Informe sobre la ley agraria* (1968 edn) Edima, Barcelona, 224 pp.

Kalela, E.K. (1960) 'Orchard' versus 'naturalistic' silviculture. *Proceedings V World Forestry Congress* (Seattle): 408–10.

Klein, J. (1920) *The Mesta: a study in Spanish economic history.* Harvard University Press, Cambridge, Mass.

Kozdon, P. (1954) Los bosques de pinos insignes y su transformación en bosques mixtos, *Montes* 10(55): 57–60.

Laffitte, F. (1919) *La repoblación forestal en Guipuzcoa.* Consejo Provincial de Agrcilura y Ganageria, San Sebastian, 115 pp.

Leibundgut, H. (1960) 'Orchard' versus 'naturalistic' silviculture. *Proceedings V World Forest Congress* (Seattle), pp. 404–08.

Lillo, J. (1931) Nacionalización de las industrias de la celulosa, *Montes e Industria* 12: 297–301.

Lleó Silvestre, A. (1929) *Las realidades: las posibilidades y las necesidades forestales de Espana.* Estudios politicos, sociales y económicos, Publicación No. 6, Madrid.

Martínez Hermosilla, P. (1942) *La desamortización en la politica forestal.* Law dissertation. ETSIM, 185 pp.

Martínez Hermosilla, P. (1947) La industrialización forestal. Sus problemas actuales, *Montes* 3(16): 371–78.

Mateo-Sagasta, J. (1979) La evolución del precio de la madera, *Montes* 35(194): 367–70.

Michel, M. (1986) Semillas y plantas forestales, *Montes* (nueva serie) 9: 43–53.

Muñoz Goyannes, G. (1958) Las importaciones de maderas de roble y haya en España, *Montes* 14(80): 128.

Nájera, F. (1962) *Las maderas de crecimiento rapido en la expansión industrial de Espana.* Ministerio Agricultura, Madrid, 72 pp.

Naredo, J.M. and Marquez, J. (1987) Tentativa de evaluación económica de las repoblaciones forestales realizadas por el Estado (1940-1983). Unpublished, 46 pp.

Olazabal, L. (1856) 'Suelo, clima, cultivo agrario y forestal de la Provincia de Vizcaya', in *Cuarenta años de propaganda forestal.* Ricardo Rojas (1898), Madrid, pp. 1–121.

Olazabal, L. (1883) 'Sobre la desamortización de los montes públicos proyectada por el Señor Camacho', in *Cuarenta años de propaganda forestal.* Ricardo Rojas (1898), Madrid, pp. 441–69.

Pablo, C.T.L., Diaz Pineda, F., Martin, P. and Ugarte, F. (1991) Perdida de suelos y explotación forestal en el País Vasco, *Bizia* 6: 35–39.

Pascual, A. (1870) Sistemas forestales, *Revista Forestal* 3: 49–59, 97–110, 145–62.

Piera, A. (1985) Replies to questionnaire returned by members of the Comité de Participación Pública convenido por la CIMA. May 1985.

Ramos Figueras, J.L. (1958) Conservación de los montes españoles, *Montes* 14(84): 499–500.

Reyes, E. (1917) Intervención del Estado en los montes particulares, *Revista de Montes* 41(971): 463–71.

Robles Trueba, S. (1932) La repoblación forestal de España. El verdadero argumento, *Montes e Industria* 20: 526–28.

Rojas, J.M. (1972) Consideraciones sobre el cultivo intensivo del chopo, *Montes* 28(165): 197–209.

Ruiz Tapiador, J.M. (1965) Aumento de productividad y mejora en los trabajos de repoblación mediante la mecanización en la preparación y plantación de fajas, *Montes* 21(122): 121–24.

Senador, J. (1933) 'El problema de las repoblaciones forestales', in Asoc. Ing. Montes, *Aportaciones a la politica forestal de España*, 53–92. Rivadeneyra Sa, Madrid, pp. 53–92.

Smith, J. (1990) *The timber and wood product industries in the single market.* Club de Bruxelles, Brussels.

Tejerina, M. (1958) Necesidad de coordinar la agricultura, la ganaderia y la riqueza forestal, *Montes* 14(83): 411–14.

Urteaga, L. (1987) *La tierra esquilmada.* Serbal, Barcelona, 221 pp.

Ventallo, P.A. (1908) *La repoblación forestal y el eucalipto en ella.* Tarrasa, 341 pp.

Villacampa, A. (1868) *La reconstitución de montes, es problema vital para España.* Ed. Ibérica, Madrid, 63 pp.

# 7 Afforestation in Algeria

*Philip Stewart*

## The setting

The world is, alas, rarely without a martyr country. The place in the news that is occupied today by Croatia and Bosnia, and yesterday by Kurdistan, was once for seven long years occupied by Algeria. Along with much else that perished in the flames of the war of independence were hundreds of millions of trees. The post-independence government, which enjoyed the continuing attention of the world's press, attracted much sympathy with its grand projects for replacing the lost forests. Less well known are the remarkable ideas in reforestation developed before independence.

Algeria was given its present shape by the French, who in 1830 occupied the central part of temperate north-west Africa and later joined to it a wider portion of the adjoining Sahara. The following account is concerned only with northern Algeria, which alone has climatic conditions suitable for forests. This is a broad belt of land, varying between 250 and 400 km in width, dominated by two main ranges of mountains that run from Morocco to Tunisia (Figure 7.1). The altitude and the steepness of relief are higher on the northern ranges than on the older southern ranges, and in general terms they increase from west to east, with the highest peaks rising to over 2000 m. Between the northern and southern ranges stretch the high plains, where huge inland drainage basins have been filled with material eroded from the mountains. The western plains are separated from the eastern by the Hodna Mountains, which run from north-west to south-east, linking the coastal ranges with the Aurès Mountains.

The pattern of rainfall is controlled by the topography, ranging from 250 mm a year on the western high plains to more than 1500 mm at places in the mountains of the north-east. Except near the coast, the country is subject to frequent frosts in winter. With a regular summer drought, the area has a typical Mediterranean climate, making it biogeographically part of Europe rather than Africa. The Straits of Gibraltar formed only a couple of million years ago, by which time all the major genera of plants and animals had made their way across in both directions.

The bioclimates of North Africa are generally classified according to a scheme developed by Emberger (1955) and based on an index which amounts to $P/M$, where $P$ is the annual rainfall in millimetres and $M$ is the average in degrees Celsius of the daily maximum temperatures in July (Stewart 1969; 1974). Plotting these values against $m$, the average of the daily minimum temperatures in January, one obtains a diagram of climates that has stood up well to the test of time (Figure 7.2). Emberger distinguished four zones – humid, sub-humid, semi-arid and arid – each of them divided according to the severity of the winter into

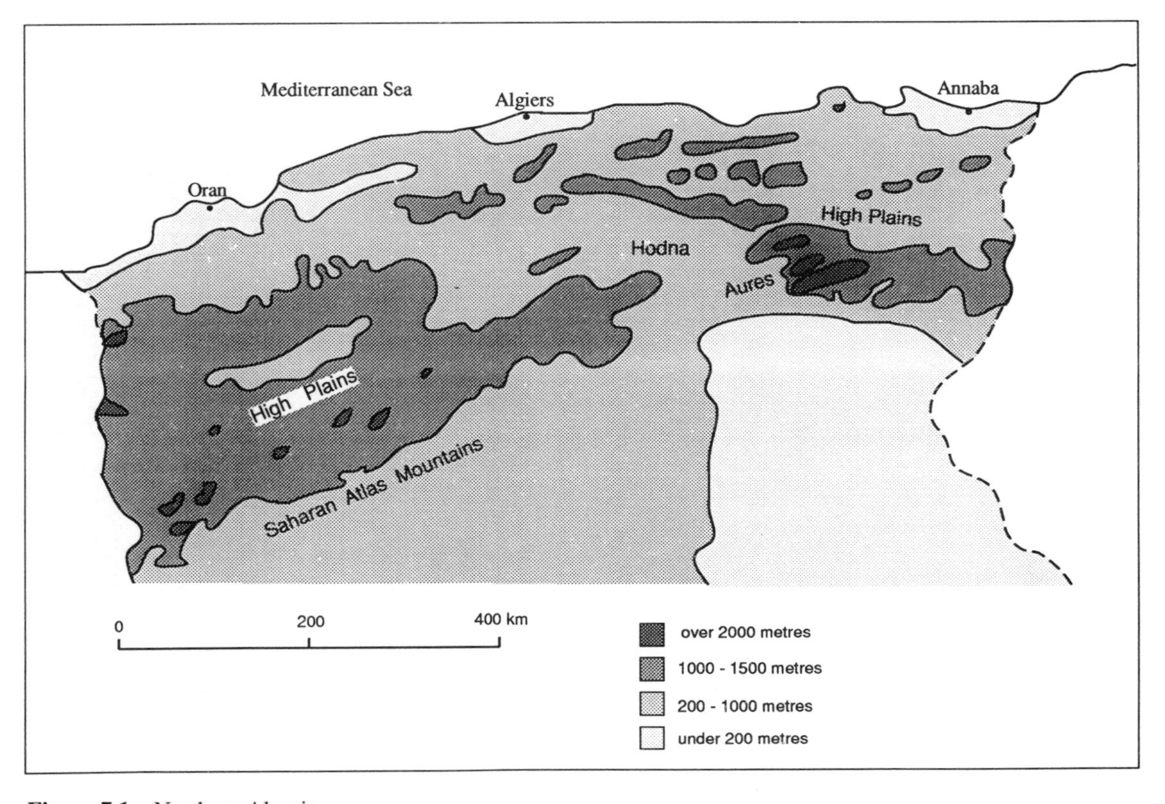

**Figure 7.1** Northern Algeria

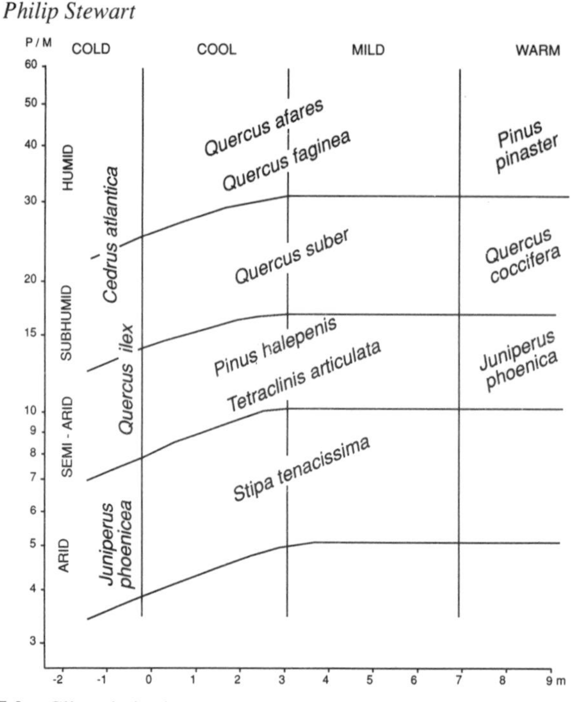


**Figure 7.2** Climatic indices and natural vegetation types.

four varieties – cold, cool, mild and warm.

The climax vegetation of northern Algeria is generally oak forest, with Zeen oak, *Quercus faginea*, and the endemic Afares oak, *Q. afares*, dominating on mild or warm humid or sub-humid sites, and with the small Kermes oak, *Q. coccifera*, locally important on old coastal sand dunes. In the colder and drier parts of the country, Holm oak, *Q. ilex*, predominates. Zeen oak has been generally replaced on acid soils in the sub-humid zone by Cork oak, *Q. suber*, which is more resistant to fire and grazing. Cork oak has itself been replaced by Aleppo pine, *Pinus halepensis*, in the mild or warm semi-arid zone, as has Holm oak on cooler semi-arid sites. Even the arid high plains, apparently already devoid of trees in Roman times, probably once had an open population of Aleppo pine, for there seems no other way to explain how they were colonised by Esparto grass, *Stipa tenacissima*, which at present establishes seedlings only under forest cover.

Humans have been present in northern Algeria for tens of thousands of years, and there are numerous 'snail-heaps' in which broken snail shells, fragments of bone and charred earth testify to prolonged occupation by palaeolithic hunter-gatherers. Agriculture arrived relatively late, probably from Egypt, but it was greatly expanded by the Romans, for whom North Africa was a granary. The eastern high plains in particular – lower-lying and better watered than those of the west – were a rich agricultural region, as witnessed by the remains of many settlements and irrigation networks. The wealth of the Roman farms owed much, however, to the forests on the mountains above, which regulated the flow of water into reservoirs, cisterns and canals. The wildness of these forests may be judged by the fact that there were herds of elephants in them for much of the

Roman period, until they were exterminated by over-exploitation and destruction of habitat (Gsell 1913). However, the most ominous event to occur under the Romans was the introduction of the camel (Gautier 1937).

With the collapse of the Roman Empire, the population declined and farming regressed. It seems probable that forests reclaimed some abandoned farmland at this time. Al-Nuweiri, quoted by the historian Ibn Khaldun (1332–1406) claimed that, at the end of the seventh century, trees (not necessarily forests) shaded the road all the way from Tripoli to Tangiers, and there was no doubt some truth behind this, even allowing for literary exaggeration. The immediate effects of the Arab invasions on the ecology of Algeria were probably less important than often supposed, though in the long term the weakening of administration gave harmful added freedom to nomadic graziers.

The first reliable scientific description of the forests was given by the French in the decades following their occupation of Algeria in 1830. However, the 132 years of French dominance were a period of rapid change, with two factors devastating Algeria's forests: population increase, from between 1 and 2 million to 10 million, and, in the last eight years of French rule, warfare. The area of forest, measured as 5 million ha in the early years of French settlement, had been reduced to 2 million ha by the time of the last nation-wide inventory in 1935, of which only an estimated 1 million ha remained in 1962. Most of the 3 million ha lost between 1830 and 1935 had been classified as communal forest and left to the communes to manage for firewood and grazing. The remaining 2 million ha were mostly state forest, except for about one-fifth which was Cork oak forest privately owned by European settlers. Some deforestation took place during World War II, to provide emergency supplies of timber and fuelwood, but most of the million hectares lost since 1935 were destroyed during the war of independence. The nationalists took to the *maquis,* forest guards were withdrawn and the army attacked by every means including napalm.

The effects of population increase were exacerbated by the fact that the best agricultural land was reserved for settlers, forcing the indigenous people to choose between being landless farm labourers on French-owned farms and seeking land for themselves in the mountains. Over-population was particularly heavy in the areas inhabited by the Berbers, who had already previously retreated into highland fastnesses to escape from control by the Arabs. Forests were eaten away to make room for traditional farming methods, alternating cereal and fallow, combined with free-range grazing, on slopes too steep to resist erosion. Demand for firewood and charcoal, especially from the three evergreen oaks (Holm, Cork and Kermes), was equally intense. Forest fires, easily started accidentally or deliberately in the summer drought, greatly speeded the devastation.

## Reforestation

### *Introduction*

The role of the State Forest Service changed through time. Until the 1870s or 1880s, the main task had been to survey, delimit and map the forest areas. For the next half century the essential concern was with guarding and exploiting the state-owned forests, though problems of regeneration grew steadily more pressing. In the 1920s and 1930s it came to be seen that natural regeneration could not be relied on to maintain the forest area, and there were increasing efforts at establishing forest plantations. At the same time, the dangers of erosion were recognised, and a 'Service for the Defence and Restoration of Soils' was set up, largely by the efforts of members of the Forest Service, to which it became administratively attached after a period of autonomy.

Three kinds of tree accounted for most of the planting in the early days: Cork oak, Aleppo pine and eucalypts. The interest in Cork oak stemmed from the fact that it was economically by far the most valuable indigenous tree. This species of oak is confined to the west Mediterranean area, from Portugal and Morocco to Italy and Tunisia. Algeria was second only to Portugal in the extent of its Cork oak forests and in the quality of their product. Unfortunately, the very cork that made for the species' biological success by conferring extraordinary fire resistance attracted the interest that made for its degradation. Repeated exploitation of the outer layer of cork reduced the trees' vitality, exposed them to sun scorch and destroyed their fire-hardiness; damage during harvesting killed some trees and allowed the entry of pests and diseases into others; and the increased human activity brought by management and harvesting virtually put an end to natural regeneration. For a plantation species growth was very slow – about one cubic metre per hectare per year, and the first harvest of cork could not be made before a tree was 40 or 50 years old.

Aleppo pine was next to Cork oak in economic importance. During the first century of French colonisation it had been exploited for its turpentine as well as for its timber. With its abundant seedlings and its ability to colonise bare land, it was alone among native species in being able to hold its own in the face of human pressures. The other valuable native conifer, Atlas cedar (*Cedrus atlantica*), was, on the other hand, notoriously difficult to regenerate, seeming to depend on a complex interaction with Holm oak. The growth rate of both these conifers was about as slow as that of Cork oak.

Eucalypts were extensively planted as specimen trees by the French settlers. More than a hundred species were found when a systematic survey was undertaken. Foresters were excited by their remarkable growth rates (Monjauze 1958). The most widely successful in the humid and sub-humid north-west was *Eucalyptus camaldulensis*, closely followed by *E. globulus* and *E. gomphocephala*. In the warm semi-arid north-west, *E. cladocalyx* and *E. sideroxylon* were impressive.

The history of afforestation in Algeria can be divided into a number of periods. Before 1939 most attempts were made with Aleppo pine, which succeeded, and with Cork oak, which failed. The period from 1940 to 1944 was an interval imposed by the war. From 1945 to 1959 was an experimental period, seeing the establishment of many species trials, especially with eucalypts, and the development of new techniques of planting. The results of all these experiments were brought together in the revolutionary 'Rural Renovation' programme of 1960 to mid-1962. After 1962 there was a return to the safe old reliance on Aleppo pine, but since 1974 some interest has been added by the Green Dam Project.

The most exciting of these periods, which gives Algerian tree-planting its uniqueness, is the shortest (1960–62) but it needs to be seen in the context of what happened between 1945 and 1959. The whole of this epoch was dominated by the ideas and realisations of one remarkable man, Alexis M. Monjauze, whose career in the Algerian Forest Service spanned more than 40 years (1929–70) and who was director in the crucial years between 1960 and 1962.

## *1945–59*

The first innovation of the post-1945 period was the establishment of a number of arboreta, most notably those of Meurdja (originally started in 1930) and Bainem near Algiers, Ain Oussera on the high plains, and Hakou Feraoun, Tamelaka and Tenira in the north-west. Though called 'arboreta', they were in effect experimental plantations, with up to a hectare devoted to each stand. They have been criticised for the over-emphasis on eucalypts and for the lack of replication; but no one foresaw the disaster that was to overcome the Australian genus, and the experimental design was constrained by budgetary problems and by the difficulty of protecting the sites against pressure from neighbouring farmers and graziers.

The second field of experiment was in techniques of soil restoration and erosion control. The most spectacular was the use of heavy rippers to remove a surface crust and break up a compacted soil to a depth of up to 80 cm (Monjauze 1954). A variant invented in Algeria – by a bulldozer-maintenance mechanic – was the *rasette*, in which the central teeth of the ripper were replaced by a horizontal blade attached to the two outer blades. Using such machines, it was possible to plant trees on sites where they would not otherwise survive, and to achieve in a single operation the revival of the soil that would otherwise need the biological work of decades.

Using another technique, invented by Monjauze, the 'steppe method', it was possible to plant trees even in the arid zone: after the passage of the ripper, the loose surface soil was piled into ridges 50 cm high, on the crest of which trees were planted (Monjauze 1960a). The essential idea was to concentrate the storage of rainwater on a fraction of the area, giving the trees an effective rainfall of twice or more the actual value. Evaporation was spread over more of the year, raising air humidity and lowering temperatures. Another method of ensuring success on very dry sites, also invented by Monjauze, was the 'masked pit' (*potet masqué*), which involved planting a tree in a sunken chamber, closed above by flat stones, leaving just room for the stem to emerge. Separately or in combination, these two techniques made planting possible even on arid sites on the high plains.

However, the ideas of Monjauze extended far beyond technical questions. He was one of the first people anywhere to see that the problems of forestry in poor countries are above all human problems (Monjauze 1960b). Most of the damage to existing forests came from the desperate search of a rapidly growing population for land on which to produce food. Because the techniques of food production had not changed, they became increasingly destructive as they were practised more intensively and pushed on to more unsuitable sites. The forester's traditional solution of using repressive methods to keep people out of the forests simply made matters worse by adding to the hostility of the population and making their co-operation impossible to obtain. The forest problem could not be solved without a solution to the agricultural problem.

To put things in historical perspective, it should be noted that Monjauze was developing these ideas at the time when the Food and Agriculture Organisation was preparing its disastrous manifesto on 'The role of forest industries in the attack on economic underdevelopment' (FAO 1962), which sanctioned the divorce of industrial forestry from the solution of other rural problems. Whether because his ideas were too far ahead of his time, or because they emerged from a colonial situation, they found no echo in the rest of the world. In Algeria, however, they came precisely at the moment when the government was ready to hear them.

### *1960–62*

The Algerian War was in origin a peasants' revolt. It began in the Aurès Mountains and rapidly spread to Kabylia – both Berber-speaking areas – and its strength lay in its control of the *maquis* (scrub vegetation that forms a dense under-storey in degraded forests on acid soils, and replaces them where they have been destroyed). Having accorded the non-settler countryside a very low priority for more than a century, the French government became aware that its grievances were real and deep-seated. The last, desperate attempt to bring prosperity to the poor of Algeria was the Constantine Plan, announced in 1959: a massive programme of investment in every sector of the Algerian economy. Officials who had been calling in vain over many years for action suddenly found that vast resources were promised to them. Among the best prepared was Monjauze, whose grand design for rural renewal already existed in outline (Monjauze, unpublished papers). Under his leadership a survey of the rural situation was carried out by early in the following year of 1960.

The 1960 survey was founded on a land classification based on rainfall, slope and soil condition. Of the 28 million ha of northern Algeria, only two-fifths were found to receive a rainfall greater than 400 mm a year – the minimum for rain-fed cultivation – and of this well-watered area only 19 per cent, mostly settler-occupied, had a slope of less than 1 in 30 – the maximum for cultivation without erosion-control measures. Out of 5 million rural Algerians, some two-thirds were living on the 52 per cent of this zone with slopes of greater than 1 in 8. In these densely populated hills, good, deep soil covered only 3 per cent of the area, while rough grazing, forest or *maquis* occupied 64 per cent. Most of the remaining third of rural Algerians were living on land that was relatively flat but which had only a semi-arid climate – the eastern high plains and the northern edge of the western high plains. In this marginal zone cereal yields were locally quite good, but grazing of sheep and goats played an important part.

A survey of land ownership other than by French settlers showed that out of just under 600,000 holdings, 72 per cent were of less than 10 ha (considered the minimum necessary to provide subsistence for a family), with 53 per cent less than 5 ha. The majority of families – three-quarters of a million – had no land at all. These figures were even worse than they seem at first sight, for extensive unproductive land on farms was included in the area covered. Redistribution of the existing agricultural land (excluding settler farms) would have provided 4.4 ha per family, with no allowance for population increase.

The Rural Renovation Programme was founded on the idea of expanding the area of cultivated land largely at the expense of degraded pasture, especially in the semi-arid zone. This was to be achieved by revolutionary new methods of soil restoration. The chief of these was the extension to agricultural land of the soil renovation by ripper that had proved so effective in reforestation.

After soil restoration was to come the construction of contour earthworks, to cut short the accelerating downward flow of water and canalise it towards specially reinforced water courses. For reforestation or orchards small earthworks close together are the best solution and can be built by hand. For pasture or farmland on gentler slopes, wider spacing and hence larger earthworks are preferable, to allow greater freedom of movement, but there is a maximum possible distance between them, beyond which erosion sets in, and by the usual standard they would still be inconveniently close together. Alternatively the earthworks may be given such a shallow cross-section that they may be ploughed along and sown with crops, but in this case they are very fragile and are destroyed by slight errors of farming technique. The new solution adopted by Monjauze in 1960 was to build extra-large and strong earthworks at extra-wide spacing. Between them there were to be secondary obstacles to erosion so narrow as to be small hindrance to agriculture: lines of stones, strips of permanent fallow, or perhaps hedges (Gréco 1966; 1978).

This investment in the infrastructure was seen as only the first part of a programme that must include changes in crops and methods of cultivation, in animal husbandry, and above all in education, habitat and social services. The authors of the programme realised that if it was to succeed it must change not just the physical base but the whole way of life. Changes in agricultural practice were of particular importance, as it was the intensification of traditional methods that had led to so much soil degradation and erosion. In particular, it was proposed that free-range grazing should be eliminated. At the time there was unrestricted access for herds to two-thirds of the area of northern Algeria and restricted access to most other land not currently under crops. Even under the new dispensation, grazing land was to be reduced by only a third, but it was to be managed pasture, reseeded where possible, with animal numbers and movements carefully monitored and controlled.

The main improvement was to be an increase in the area under crops and orchards to 9.3 million ha from 3.8 million (excluding fallow), thanks partly to soil restoration, partly to the planting of fodder trees, partly to the use of special procedures to bring more semi-arid land under culture. The reduction of the area of pasture was to be compensated by a very great increase in the amount of fodder produced and of animals raised. It was hoped that the population could be persuaded to abandon freedom of grazing in return for help with setting up intensive animal-production units.

One point of special significance was the prediction that by the end of the programme, it would be possible to maintain 20 million sheep, as against the ceiling of 6 or 7 million sheep and goats under the old system. Fodder for the winter and the dry season was to come partly from annual crops but still more from a planned 1.7 million ha of fodder orchards (*vergers semi-forestiers*) – plantations of tree species such as carob (*Ceratonia siliqua*), and honey locust (*Gleditschia triacanthos*), and of prickly-pear cactuses (*Opuntia* spp.) (Lavallée *c.* 1962; Monjauze and Le Houérou 1965). The advantage of these was that they could occupy land unsuitable for food crops or for pasture. The produce of the fodder orchards was not to be for direct consumption but was to be processed industrially with simple technology, making possible bulk-harvesting systems in which it would not be necessary to separate out twigs or thorns.

A great expansion was envisaged also in conventional orchards to 2 million ha, for the production not only of established Algerian fruit such as figs and olives but also of new fruit crops with export potential. Pistachio nuts were seen as an especially promising line, for the native *Pistacia atlantica* was widespread and could serve as a rootstock for grafting (Monjauze 1965). The Berber-speaking peoples, in particular, had a long tradition of cultivating fruit trees and were likely to take advantage of the opportunity to develop it further.

Great flexibility was envisaged in the integration of this tree production with agriculture and forestry. At one extreme, some fodder trees were to be planted as parcels or belts within forests, for example around foresters' houses or in firebreaks. At the other extreme, fruit or fodder trees were to stand on the bench-terraces between fields of annual crops. Here were the ideas that were later to be labelled 'agro-forestry', 20 years before the invention of that deplorable word (Stewart 1981).

It was hoped that the growing prosperity of the rural population and the end of free-range grazing would make possible a renewal of the forests, with a target of 1.5 million ha of production forest and 1.6 million of protection forest. Together with the fruit and fodder orchards, this would bring the total area under trees to 6.8 million ha, more than in 1830.

Most intangible, but perhaps most interesting, would be the effects of the new landscape on people's ideas. In the place of a jumble of small fields encircled by wasteland, the eye would take in long strips of different crops filling all the gentler slopes, and blocks of forest or orchard or managed pasture on the steeper ones. In the place of downward-striking ravines, the view would be traversed by the curving horizontal lines of trees, hedges and roads running along contour earthworks. Living in an orderly, stable and fully used setting, the people would perhaps come to face nature with hope and confidence instead of seeing in it hostile and capricious forces.

The programme as a whole was never officially adopted, nor was the necessary legislation ever enacted, though many elements found their way into the law governing the Agrarian Revolution of 1971. As a country-wide plan it was soon forgotten, leaving many traces in the form of statistics and techniques and, above all, of pilot zones where the main work of soil restoration and erosion control was more or less completed and where some at least of the tree-planting was carried out. There were originally six such zones, with a seventh added after independence. They were all situated on marginal land with rainfall of about 400 mm per year, soils impoverished by erosion and slopes ranging from moderate to steep. The principal criterion of selection was one of demonstration value: the zones treated were to convince both the people and the authorities in each region that Rural Renovation worked, even on the least promising land.

These pilot areas were known as ZORs (*Zones d'Organisation Rurale*) – an unfortunate choice of term, as *zorr* means 'compulsion' in Arabic. They ranged in area from 10,000 to 20,000 ha. The essential novelty of the ZOR, compared with any previous attempt at rural development, lay in a number of features:

integration, all factors of production, including the soil, being acted on together; urgency, the project being carried out as fast as possible, so that the upheaval would be brief and benefits early; and scale, each ZOR being large enough to extend over a whole catchment area so that the flow of water could be regulated throughout a system of erosive water-courses, providing a full range of land uses for a whole community.

In each pilot area an intensive campaign of several months informed the inhabitants of the coming work and sought their co-operation. The project was then carried out as fast as possible. The owners were compensated for any lost crops, and their landmarks showing the boundaries of their property were removed and replaced in their presence before and after treatment. It is hard to know how peacefully this process was carried out in the prevailing war-time conditions. The most convincing evidence of acceptance is the fact that after independence, the leaders of the commune neighbouring the ZOR of Zeriba asked for soil renovation to be extended to their land.

After independence, work continued at a reduced rate at two of the ZORs. It appears that President Ahmad Ben Bella was enthusiastic for Rural Renovation. Activity was halted after his overthrow in 1965, when the new President, Houari Boumedienne, decided to interrupt all projects considered non-essential, pending their review. Work at Zeriba was nominally restarted in 1972, though with little of the original vision.

Various criticisms of the Rural Renovation Programme have been made retrospectively, but the greatest technical complaint has always centred on the main earthworks. Part of the problem is that the secondary obstacles were not built, and without them erosion was not completely stopped. Another mistake was that they were often made with topsoil instead of poorer material from deeper levels. But most criticism has been made of their sheer size, which was often more than was strictly necessary for the probable maximum flow of water. The fact remains that, alone of French erosion-control works in Algeria, after 30 years they are still largely intact.

### *1962–74*

Independence in 1962 brought the exodus of hundreds of thousands of French residents, including almost all the administrative staff of the Forest Service and of the agricultural services. Monjauze remained for several years as adviser to the Minister of Agriculture, and a much reduced programme of Rural Renovation continued until 1965; but the essential concern now was simple survival in face of the lack of skilled personnel and the huge task of reorganizing and running the farms abandoned by the settlers.

The situation was particularly bleak in the forests, unguarded and unmanaged since the mid-1950s, full of gaps and with a chaotic distribution of age classes following all the forest fires. In many districts, the National Liberation Front had made rash promises of free access to forests, to win popularity with villagers. Worst of all was the outlook for the Cork oak forests, unregenerated for decades and now bringing in a fraction of their former revenue. They no longer had the economic value to compel government attention (Monjauze 1965–66).

In effect, Algeria had to start a new forest service, with a handful of fresh graduates from European forestry schools and with several hundred technical agents who were given a basic crash-course. These raw recruits were sent to posts where there were no old hands to complete their education in the field, and where shortages of essential equipment would have taxed the ingenuity even of experienced people. Many soon found other ways of making a living. However, the government expected a great deal from this ragged band, for it had adopted an ambitious reforestation target of 60,000 ha a year as one of its key policies for the countryside.

The outcome was a return to the old practice of planting almost nothing but Aleppo pine, a species so tolerant, so plastic and so unappetising even to goats that it succeeded on most sites. No sensible young forester was going to risk failure with something difficult when there was such a safe alternative. In the end this prudence was to prove well founded, for a major pest, *Phoracantha semipunctata*, was to devastate the country's eucalypt plantations in the 1980s. There were failures even with Aleppo pine, but in the absence of adequate controls it is impossible to say what the actual gross rate of afforestation was.

In reality, in spite of the bold claims made, the net rate of reforestation was probably near to zero or even negative. Forest fires were each year destroying an area of forest of the same order of magnitude as that which was planted, and the search for cropland and pasture continued to make quiet inroads. The planting was at best a holding operation. The Forest Service that had so recently been ready to lead a rural revolution was now one of the feeblest organisations in the new republic. It no longer even carried out its own projects, for that function was transferred to a new enterprise – the 'Office National'.

### *1974 to the present*

The latest period is one of the most bizarre in the history of world forestry. What had been lacking since 1962 was any sort of grand design. The gap was filled in 1974 with the adoption of a gigantic scheme for a 'green dam to hold back the Sahara'. A belt of trees 20 or 30 km wide was to be planted on the southern face of the Saharan Atlas, Hodna and Aurès Mountains, stretching all the way from the Moroccan to the Tunisian frontier – a distance of about a thousand kilometres. The aim was to afforest up to 3 million ha, using several billion trees.

The most bizarre aspect of this whole scheme was its genesis. The origin can be traced directly back to none other than St Barbe Baker, founder of 'The Men of the Trees'. In a long life of ceaseless effort he failed to persuade any government or international institution to act on his ideas of greening the world's deserts. However, one of his disciples was to have more success: a New Zealander called Wendy Campbell-Purdy. Though without any background in forestry, she was persuaded in the course of a day's inspiring conversation with Baker to devote the rest of her life to his vision.

After false starts in Morocco and Tunisia, Campbell-Purdy arrived in Algeria in 1962, just when everything was possible in the administrative chaos that followed independence. She persuaded the local authorities to give her the use of several hundred hectares of land, just north of the town of Bou Saada; she obtained plants – mainly eucalypts and acacias – from Forest Service nurseries; and she was given money, dried milk and old clothes by War on Want to pay her planters (Campbell-Purdy 1967). By 1970 her plot of land was covered with trees. It was a real achievement, though it must be noted that the land lay at the mouth of a wadi and received much more water than the 200 mm annual rainfall; the trees were also irrigated with sewage from the town's cesspits, which added to their vigour as much as it detracted from their charm.

Unfortunately, Campbell-Purdy was no administrator. She left the management of the project to her Algerian right-hand man, Aissa, an energetic young man of doubtful honesty. Living as a solitary western woman surrounded by traditional Islamic society, she sank into a depressed state, leaving all the running of the project to Aissa, and was finally persuaded to leave. Faced with the evidence of serious mismanagement, War on Want withdrew its support and the project collapsed. Campbell-Purdy made half-hearted attempts to start again in other countries, but her health never recovered and she died in Greece in 1985, aged only 59. She never went back to Algeria, and the authorities there never acknowledged her influence, but it seems virtually certain that the Green Dam Project owes its origin to her. I understand that it was one of her Algerian associates, an ex-minister working for his rehabilitation, who was the hidden persuader.

The Green Dam Project is spectacular, but it is intellectually a feeble successor to Rural Renovation. It is based on a fallacy – that the desert is something that creeps forward at ground level, and can be halted by a barrier of trees. In reality, desertification is a process that takes place wherever the necessary conditions prevail. The climatic effects of the Green Dam will not be felt more than a few kilometres away, and northern Algeria will go on turning to desert wherever over-exploitation and deforestation continue to destroy the soil.

Even leaving aside the fallacious primary objective, the project is misguided. The area is sparsely inhabited and in the absence of local labour the work is being carried out by the army (and with minimum reference to the old Forest Service). No interaction with food production is in view, and the plantations will not begin to produce wood in useful quantities for several decades, so the economic benefits will be small and long delayed. Indeed, there is no part of Algeria where the planting of 3 million ha of forest could provide fewer benefits to society.

Perhaps the saddest aspect is that the plantations of the Green Dam are only a short distance from the last surviving natural forests of the southern mountain ranges, where hundreds of thousands of hectares of Holm oak, Aleppo pine, Juniper (*Juniperus phoenicea*, *J. oxycedrus* and *J. thurifera*) and, in places, Cedar are vanishing under the combined effects of illicit grazing and woodcutting and of forest fires. By drawing all the attention and resources away from what already exists, the new plantations prevent the undertaking of what should be the most urgent task, that of conservation.

## Conclusion

The reforestation that drew attention to Algeria was that which followed independence, but this was in fact far less interesting than the ideas and techniques of the 1950s. The world has still not caught up with the authors of Rural Renovation, who were among the first to see that the forests will not be saved if people cannot feed themselves. Until this notion is generally accepted, the martyrdom of the forests will continue, not just in Algeria but in all the hungry countries of the world.

## References

Campbell-Purdy, W. (1967) *Woman against the desert*. Gollancz, London.

Emberger, L. (1955) Une classification biogéographique des climats, *Recueil de Travaux, Laboratoire de Botanique, Géologie, Zoologie, Faculté de Sciences de Montpellier, Série botanique*, no. 7, Montpellier.

FAO (Food and Agriculture Organisation) (1962) 'The role of forest industries in the attack on economic underdevelopment' in FAO, *The state of food and agriculture 1962*, FAO, Rome, pp 88–128.

Gautier, E.F. (1937) *Le passé de l'Afrique du nord*. Payot, Paris.

Gsell, St. (1913) *Histoire ancienne de l'Afrique du nord*. Hachette, Paris.

Gréco, J. (1966) *L'érosion, la défense et la restauration des sols, le reboisement en Algérie*. Ministère de l'Agriculture et de la Réforme Agraire, Algiers.

Gréco, J. (1978) *La défense des sols contre l'erosion*. La Maison Rustique, Paris.

Lavallée, P. (*c*. 1962) *Le caroubier: son utilisation dans l'alimentation du bétail en Algérie et en Tunisie*. Direction de l'Agricuture et des Foràts, Algiers.

Monjauze, A.M. (1954) *L'emploi des rooters dans le reboisement*. Report of the Groupe de Travail Eucalyptus. Station de Recherche Forestière du Maroc, Rabat.

Monjauze, A.M. (1958) Note sur le développement des eucalyptus dans certains arborétums de l'Oranie semi-aride froid, *Bulletin de la Société d'Histoire Naturelle de l'Afrique du Nord* 49: 143–60.

Monjauze, A.M. (1960a) Le reboisement sur rootage en plein et sur bourrelets, *Revue Forestière Française* 12: 1–25.

Monjauze, A.M. (1960b) 'Essai pour l'utilisation rationnelle des terres en zone aride et semi-aride'. (Commissioned but not published by FAO; internally circulated in Délégation Générale du Gouvernement en Algérie).

Monjauze, A.M. (1965) Répartition et ecologie de *Pistacia atlantica* en Algérie, *Bulletin de la Société d'Histoire Naturelle de l'Afrique du Nord* 56: 5–128.

Monjauze, A.M. (1965–66) Le chêne-liège: peut-on rénover la forêt?, *L'Algérie Agricole* 7: 30–36; 8: 65–75.

Monjauze, A.M. and Le Houérou, H.N. (1965) Le rôle des *Opuntia* dans l'economie agricole nord africaine, *Bulletin de l'Ecole Nationale Supérieure d'Agriculture de Tunis* 8/9: 85–164.

Stewart, P.J. (1969) Quotient pluviothermique et dégradation biosphérique: quelques réflexions, *Bulletin de la Sociètè d'Histoire Naturelle de l'Afrique du Nord* 59: 23–36.

Stewart, P.J. (1974) Un nouveau climagramme pour l'Algérie, *Bulletin de la Sociètè d'Histoire Naturelle de l'Afrique du Nord* 65: 239–52.

Stewart, P.J. (1981) Forestry, agriculture and land husbandry, *Commonwealth Forestry Review* 60: 29–34.

1 Private-sector afforestation in Glen Spean, Scotland. Most of the forest, which was planted at the beginning of the 1980s, is of Sitka spruce. The land was previously used for hill-sheep farming.

2 Forestry Commission afforestation on steep slopes overlooking Loch Leven in west Scotland.

3 Land prepared for forest planting, near Achnasheen, Scotland. Erosion of such land, and the accompanying silting of streams and rivers, has attracted attention in some areas, and guidelines setting out good practice have now been issued by the Forestry Commission.

4 Forestry Commission forest at Culbin Sands, Scotland. Formerly an extensive area of unstable sand dunes, Culben was extensively afforested between the two World Wars. (J. Livingston)

5 State forest blocks in County Wicklow, Republic of Ireland. This is an area of small farms and fields and more extensive hill grazings, where most of the afforestation is located.

6 Sand dunes and coastal afforestation in west Jutland, Denmark. (W. Ritchie)

7 Northern Algeria: area treated with a variety of erosion control techniques during Rural Renovation. (P. Stewart)

8 Chile: aerial view of stands of trees on Shell's Rucamanqui forest estate on the western foothills of the Andes. (Shell)

9 New Zealand: a pervasive image of the first planting boom. Workers planting trees on the Kaingaroa plain in 1922. During the Great Depression, unemployed relief work schemes enabled initial planting targets to be exceeded by about 25 per cent. By 1970, Kaingaroa had reached 122,758 ha and for a time enjoyed the reputation of being the largest single plantation in the world. (New Zealand Forest Service Collection, National Archives, Wellington)

10 Conical Hill Sawmill, Otago, photographed in 1986. This plant, the South Island equivalent of the Waipa Mill, set up in 1942 in the Bay of Plenty, was also designed to demonstrate more advanced Swedish frame saws and to pioneer the development of sawing of Radiata pine. In 1988, Conical Hill was placed on sale and was purchased by Ernslaw One Ltd and run under the name of its subsidiary, Blue Mountain Timber Company. (M. M. Roche)

12. USA: Tree planting on badly eroded land in Tennessee sometime during the 1930s. (Bob Wallace – held by Forest History Society)

---

11. *(bottom left)* USA: Cut-over land in Louisiana. After rapacious logging of the southern forest during the early years of this century logging firms usually tried to sell off the land for small-holdings. But the usual progression of forest to farmland often failed on areas of poor, sandy soils (such as this) or in marginal agricultural areas as in the Lake States. Such land reverted to forest, although often with a depauperate stand. (Forest History Society)

# 8 Afforestation in China

*Vaclav Smil*

## Introduction

The story of Chinese afforestation goes far beyond a dispassionate evaluation of the nation's advances in planting trees. In Maoist China, where nothing had escaped the overblown politicisation, annual mass tree-planting campaigns were among the leading repositories of propagandistic sloganeering and exaggerated claims of unprecedented achievements. Uncritical Western observers, enthralled by the mixture of the long-lasting fascination with the affairs of the Middle Kingdom and of the more recent awe at the great social experiment, were only too eager to believe every fabulous claim made by Beijing bureaucrats, or to describe typical experiences of a few show places as the national norm.

John K. Galbraith, glancing from the windows of Beijing–Guangdong train, concluded that 'the hills of China, which I had always heard of as being bare, are no longer so' (Galbraith 1973). And Jack Westoby, after unequivocally approving Maoist forestry policies, claimed that 'the mightiest afforestation effort the world has ever seen' means 'that the balance between man and nature, lost through centuries of reckless forest destruction, is being steadily restored' (Westoby 1979).

Increasingly realistic Chinese reporting, initiated in 1979 by Deng Xiaoping's bold post-Maoist reversal toward social and economic reforms, has provided a very different appraisal of the country's much extolled afforestation programme – but the central bureaucracy is still issuing bloated annual totals and revealing plans whose enormity guarantees their failure. In this chapter I will try to sort official exaggerations from indisputable realities, and to offer a balanced assessment of China's achievements in afforesting that badly deforested land.

## China's forestry

China's need for extensive afforestation is all too obvious. The latest survey of the country's forest cover, undertaken between 1984 and 1988, ended up with 124.653 million ha (Mha) of forests, including 86.354 Mha in mature stands and 38.299 Mha in plantations (Ministry of Forestry 1990). The total represents just 13 per cent of China's territory, and it prorates to a mere 0.11 ha per capita, or to just one-eighth of the global average. Moreover, poor stocking, averaging less than 75 $m^3$/ha, means that the total wood reserves amount to only 9 billion $m^3$ ($Gm^3$), or less than 8 $m^3$ per capita.

The total volume of commercially distributed timber has been most recently

**Figure 8.1** Chinese provinces.

about 55–60 million m$^3$ a year, in per capita terms again an order of magnitude below the world-wide mean. A good way to appreciate the relative meagreness of this use is to translate that volume into readily imaginable equivalents: 0.05 m$^3$ of wood used annually per person is a volume smaller than that of a standard interior house door with its casing – or, when all of it would be pulped, enough to produce just about 15 kg of paper. And the annual harvest of fuelwood works out at less than 3 kg of fuel a day per rural household – while the surveys show that 11 kg a day is the existential minimum (Smil 1988).

Keeping in mind the scores of critical needs for wood in modern society – from coal-mine props to computer print-outs, and from concrete forms to toilet paper, it is no surprise that wood is among China's rarest commodities. And yet much wood is wasted, and shortages of Chinese wood supply could be eased significantly by higher rates of woody biomass utilization (Long 1989). Future demands make for a very worrisome outlook. In spite of a relatively successful population-control programme, China's absolute annual increases are now in excess of 15 million people; the country's bold modernization goals, aimed at a quadrupling of the 1980 gross domestic product by the year 2000, create an unprecedented demand for all natural resources; and the highly uneven distribution of remaining forest, combined with inadequate transportation capacities, greatly aggravates regional shortages.

The area of state-owned forests shrank by almost 25 per cent (12.85 Mha) during the 1980s, with mature growth accounting for 60 per cent (7.69 Mha) of this large decline (Li 1989). As a result, 82 per cent of China's timberlands in 1990 were young or middle-aged stands. Harvestable timber reserves declined by 2.308 $Gm^3$, with mature stands contracting by 170 $Mm^3$ annually. Moreover, the growing stock ready for harvesting in mature forests now amounts to less than 1.5 $Gm^3$, and it could be all cut in just seven to eight years. Reserves approaching maturity will shrink from about 2.6 $Gm^3$ in the late 1980s to just 1.25 $Gm^3$ by the end of the century. Wood shortages have already forced China to become a major timber importer: with the exception of 1985, the country has spent more than $1 billion a year every year since 1984, mainly for shipments from Canada and USA (FAO 1991).

Deforestation has been also driven by expansion of croplands and pastures, and its ecosystemic effects are seen above all in much increased soil erosion, stream silting and higher frequency of damaging floods (Smil 1984; 1992). Officials of the Ministry of Water Resources admitted that the damage caused by deforestation in the Yangzi (Chang Jiang) basin has already far outstripped what soil-erosion control can restore (Li 1989), and by 1988 the forest cover in Sichuan, China's most populous province (Figure 8.1) and, thanks to its western mountains, one of the country's principal timber bases, was down to 12.6 per cent from 19 per cent in the early 1950s (Xinhua [China's News Agency], 30 November 1988, *Joint Publications Research Service JPRS* CAR 89–005).

## Policy developments

As with all of the country's major public policies, post-1949 developments of China's forestry have closely followed the twists of the official party line (Ross 1980; FAO 1982; Smil 1984; Ross 1988). The Agrarian Reform Law of 1950 nationalized about three-quarters of all forests, and small private woodlots (many established by redistribution during the land reform of the early 1950s) were soon taken over by new agricultural co-operatives.

By 1957, the last year of China's first Stalinist Five-Year Plan, rural collectivization was complete, and for the following 22 years China's forests were managed either by state forest farms (whose count eventually reached about 4000) or by communes (whose total number before their dissolution was about 50,000).

A fundamental change came only after 1979 when Deng Xiaoping's new reform policies introduced a rural responsibility system combining collective ownership of land and a commitment to fulfil contractual obligations (including state production targets) with private management, performance incentives and free-market sale of surplus output. Contracting in forestry was actually in the forefront of rural privatization: after tentative beginnings in 1980 and 1981 the process was greatly accelerated during the following three years.

By 1984 virtually all forested or afforestable land formerly in communal, and now in township, ownership – a total of some 50 Mha – was managed by using some forms of responsibility arrangements involving 4 million specialized households. A year later a new Forestry Law and a new set of managing regulations codified these practices and allowed for an even greater latitude of private decision-making by easing some of the compulsory requirements (Ross and Silk 1987).

At the same time there was a rapid expansion of private woodlots. Before 1979 they were barely tolerated, limited to less than 5 per cent of communes' hilly wasteland, and often condemned as the last vestiges of undesirable bourgeois ownership. The reversal came in the spring of 1980 when a State Council directive confirmed that the trees planted by the villagers will belong to them, rather than to a collective: 'Whoever afforests the land owns the trees.' By 1983 some 50 million rural households had 17 Mha of private woodlots, and in the same year they gained the right of inheritance, a measure greatly promoting rational long-term management. Before the end of the decade the total area of private lots reached about 20 Mha. Consequently, by 1990 just over 70 Mha of China's forests (almost 60 per cent) were either managed by households with contractual agreements, or were directly owned by peasant families. The remainder of about 50 Mha, including virtually all mature growth in the four key wood-producing provinces (Heilongjiang, Sichuan, Yunnan and Jilin), continues to be in the state ownership and it is managed by a network of forest bureaux.

Most of the post-1949 afforestation has been always done in collectively-owned (and now privately-managed) areas. Low labour cost of these plantings – done mostly during the slack farming periods, involving nation-wide tens of millions of peasants, and requiring only modest state investment – has been their main appeal for the state planners. Since the mid-1950s these plantings have accounted for at least 80 per cent of all newly treed areas, mostly in fast-growing species used for timber and fuelwood.

By far the most popular trees have been poplars (*Populus tremula*, *P. cathayana*, *P. davidiana*, *P. diversifolia*, *P. koreana.*, *P. simonii* and a number of other species) in northern provinces, Chinese firs (*Cunninghamia lanceolata*) in the Yangzi basin, and bamboos (*Phyllostachis*) in the south. Collective shelterbelt and windbreak plantings, especially extensive in arid northern provinces and on the North China Plain, have used mainly poplars, willows (*Salix*) and ashes (*Fraxinus*).

On an annual basis the state forest bureaux have accounted for no more that 10–20 per cent of new plantings, mainly as restoration of the prime logging areas in Heilongjiang, Jilin, Nei Monggol and several southern provinces, and as expensive reforestation (frequently involving air seeding) of remote hilly and mountainous regions in north-west (Shaanxi, Gansu) and south-west China (Sichuan, Yunnan).

The most commonly planted species include pines (*Pinus sylvestris*, *P. masoniana*, *P. koraiensis*, *P. yunnanensis*, and *P. armandii*) in all regions of China, Chinese fir (*Cunninghamia lanceolata*), fox-glove tree (*Paulownia fortunei*) and sassafras (*Sassafras tsuma*) in central, eastern and most southern provinces, and eucalypts (*Eucalyptus spp.*) and casuarinas (*Casuarina equisetifolia*) in the extreme south.

But the state record in replanting major logging areas has been rather poor, and it is leading to a resource crisis: of the 131 state forestry bureaux in the principal production zones, 25 had basically exhausted their timber by the late 1980s, 40 can harvest for only five to ten years – and by the year 2000 almost 70 per cent of China's state forestry bureaux will basically have no trees to fell (Li 1989). Among the collectively-run forest districts, 250 counties were suppliers of commercial timber in the 1950s, but only about 100 were delivering by 1990.

And yet even a casual observer of Chinese affairs could not miss a large number of recurrent claims of record afforestation rates, of nation-wide mass tree-planting campaigns, of Chinese leaders spading the ground during the national tree-planting day. How much difference have these efforts made?

## Afforestation claims and realities

Official Chinese afforestation claims leave no doubt that since 1950 the country has been engaged in the biggest revegetation effort in human history. Between 1950 and 1957 (the period of post-war economic reconstruction and the First Five-Year Plan) claims of newly afforested areas totalled 15.7 Mha, and an identical achievement was claimed for 1958–60, the years of the Great Leap Forward (Mao's irrational attempt to accelerate the course of China's economic advancement). The total for the first half of the 1960s came to 10.5 Mha, followed by 48.3 Mha between 1966 and 1976 (the decade of the Cultural Revolution, a misnomer for a period of intensified Maoist persecution and mismanagement).

Before the end of the 1970s, official afforestation claims added a further 13.8 Mha, for a grand total of 104 Mha of new plantings during the three decades between 1949 and 1979. As China's total forested area in 1949 was most likely between 85 and 95 Mha, official afforestation claims would have represented more than doubling of the country's forest cover in 30 years. An international perspective on this unprecedented achievement is best obtained by noting that 104 Mha of new plantings is an area about 20 per cent larger than all existing (open and closed) forests in western Europe.

Critical observers have been always aware of the exaggerated nature of Chinese afforestation claims – but a quantitative assessment became possible only three years after Mao's death when Deng Xiaoping launched his badly needed economic-reform policies whose corollary was a vastly increased flow of better statistical information and, above all, an astonishing sweep of criticism and publicist as well as scientific candour. The realities concerning China's forestry gradually disclosed since the late 1970s have been quite different from the Maoist propaganda, and the afforestation accomplishments have been the subject of the greatest revisionist rectification.

By far the most outrageous was the claim of 15.7 Mha of new plantings (an equivalent of all forests in France) during the three Great Leap years (1958–60) – the time, as we know now, of the greatest famine in human history resulting in nearly 30 million deaths (Ashton *et al.* 1984). Actually the Leap years were a period of massive deforestation largely driven by Mao's delusionary goals of boosting China's iron output through a mass charcoal-fuelled smelting of locally-mined ores in primitive 'backyard' furnaces.

When it resumed its publication in 1980 (after more than two decades), China's *Statistical Yearbook* carried comprehensive series of post-1949 afforestation claims which were clearly incompatible with other official statistics, and with new admissions by the Ministry of Forestry. Indeed, the first set of new official statistics, published as a press communiqué in 1979 after a hiatus of two decades, put China's forested area at 122 Mha, or 12.7 per cent of territory. This total, based on the nationwide inventory done between 1974 and 1976 (with annual updates since 1978), referred to fully stocked productive forests, that is to the growth whose canopies cover at least 30 per cent of the ground (Beijing Home Service, 14 January 1980, *Summary of World Broadcasts SWB* FE/W1067 A/5). These forests contained about 9.5 $Gm^3$ of wood, and their annual wood increment was estimated at 220 $Mm^3$. Clearly, this inventory would have come up with very different totals if the official afforestation claims were even half true.

Various Chinese sources put the 1949–50 share of forested land as low as 5 and as high as 8.6 per cent (Yue 1980). The former total is obviously too low; the latter translates to about 83 Mha.

But the minimum 1949 total of China's forest land was almost certainly even higher, most likely close to 10 per cent, or around 95 Mha. In any case, the total included an unidentified mixture of forest categories, ranging from mature high-yielding stands (mostly northern conifers in Heilongjiang and mixed growth in Sichuan and Yunnan provinces) to scrubby woodlands.

Official statistics put China's 1949–79 timber harvests at 1.01 $Gm^3$, but the actual removal was about five times larger. This large disparity is due to the fact that the reported timber harvest represents only about two-fifths of the actually felled wood, and that fuelwood, which has never appeared in official harvest statistics, accounts for about three-fifths of the total felled volume (Chen 1983). An average annual increment of at least 150 $Mm^3$ would have produced about 4.5 $Gm^3$, which means that, without any afforestation, China's forested area should have remained roughly stable. An increase to 122 Mha from 83–95 Mha in 1949 implies afforestation of 27–39 Mha, rather than the officially claimed 104 Mha.

Revised Chinese afforestation estimates are in good agreement with this approximate assessment: according to the first survey of forest resources, mass afforestation campaigns restored tree cover on 28–30 Mha ha of land during the years 1949–79, or on less than 30 per cent of the area claimed by the past official planting statistics (Ministry of Forestry 1981). Two-thirds of these successful plantings were in timber stands, about 13 per cent in bamboo groves, just over 6 per cent in shelterbelts, and a mere 2 per cent in fuelwood lots (the remainder was in stands with mixed use).

## Afforestation problems

An apparent success rate of less than one-third is not at all surprising when one appreciates common shortcomings of Chinese statistical reporting in general, and some bizarre 'afforestation' practices in particular. To begin with, a significant proportion of the claimed afforestation totals has been always a matter of completely fictional achievements: centrally allocated afforestation targets were in due time reported by local bureaucrats as fulfilled, indeed over-fulfilled, and the upper layers of the bureaucracy were content to collate the reports into stunning totals.

Many bureaucrats reported afforested areas by simply pro-rating estimated numbers of available seedlings or saplings, no matter if the young trees were actually discarded, buried in trenches, planted upside down, or even uprooted and replanted in order to boost the total claim! As Xu Dixin (1981) noted, responsible officials in many places 'reported to higher authorities every year the area and the number of trees and, having done this, considered that their task was completed'.

But, of course, huge numbers of trees were actually planted – and abandoned. Wang Jingcai (1978) cites a new folk saying explaining why in many areas people were planting trees every year, and still had no forests: 'Trees everywhere in spring, just half left by summer; no care taken in fall, all trees gone by winter.' Survival rates of many plantings were extremely low, often only a few per cent, for a number of reasons.

The principal causes included careless planting (such as bare-root saplings put into dry rocky or sandy soils), no or highly inadequate follow-up care (no watering during China's frequently prolonged dry spells, no weeding, no protection against grazing animals), and inappropriate choice of trees (planting of species unsuitable for the climate or for local soils).

This shoddy, careless work was quite satisfactory in a system whose standard criterion for afforestation success until the late 1970s was *chenqhuo*, survival on the day a tree was planted. With such a patently ridiculous approach it was easy to realize the world's most astonishing afforestation 'success'. Although the poor success rate of less than 30 per cent was clearly representative of a majority of Chinese plantings, there was also no shortage of even higher failures – and impressive successes.

The former category covers many plantings in the arid, semi-desert to desert, north-western provinces, the latter one includes above all a class of plantings which have never entered the nation-wide afforestation statistics as they do not form extensive, contiguous tree cover. With their traditional penchant for numbering things, the Chinese call these plantings 'four-besides': trees besides houses, roads, streams and villages. These plantings, commonly composed of bamboo in southern provinces and poplars and willows in the north, also include a number of oil, nut and fruit species.

The most popular choices include tung oil trees (*Aleurites fordii*) and tea oil trees (*Thea oleosa*), walnuts (*Juglans mandshurica* and *J. regia*), Chinese chestnuts (*Castanea mollissima*), apples (*Malus*), pears (*Pyrus*), Chinese dates (*Ziziphus jujuba*), persimmons (*Diospyros kaki*) and, in the southernmost provinces (mostly in Guangdong), litchis (*Litchi sinensis*).

Privatization transformed most of the Chinese countryside, and there is little doubt that since the early 1980s the quality of many new plantings and their subsequent care have been incomparably better than during the decades of Maoist mass campaigns. The new criterion of afforestation success – good prospects for future growth (*baocun*) – is still too imprecise to offer a rigorous and generally applicable yardstick, but it is obviously superior to the previous 'one-day' farce. New private contract plantings in southern provinces have been relatively most successful.

But by far the most ambitious project has been the state-directed planting of the Green Great Wall, a massive shelterbelt designed to check the advancing desertification across China's northern provinces. Official claims put the total plantings during the project's first phase, running between 1978 and 1985, at 6.05 Mha, protecting some 8 Mha of farmland from shifting sand (Xinhua, 30 May 1988, *SWB* FE/W0030 A/5). The second phase, totalling 8.46 Mha of trees, is to be completed by 1995.

The other two massive afforestation projects now underway are a 3000 km long, 10 km wide shelterbelt along the Huang He from Zhongwai county in Ningxia to Pinzhou county in Shandong, and slopeland plantings in the Chang Jiang basin, where more than 7 Mha of new forests are to protect the river's upper and middle basin from soil erosion (Xinhua, 5 June 1988, *SWB* FE/W0030 A/5). These three huge projects would enlarge China's tree cover by roughly 25 Mha before the end of the century.

But average success rates, especially in state-sponsored plantings, remain relatively low, and even false afforestation claims did not disappear during the 1980s – but now we do not have to wait two decades to learn the extent of these gross exaggerations. A survey by the Northwest Institute of Forestry in Xi'an indicated that half of the reported national afforestation claim during the mid-1980s was false, and that the survival rate of planted trees was no higher than 40 per cent (Xi'an Provincial Service, 29 April 1987, *JPRS* CAR 87–6).

And even the properly planted trees have had great difficulty surviving the combined impacts of prolonged droughts, pest infestation and fires. During the 1980s most northern provinces, as well as some parts of the south, experienced long periods of below average precipitation or serious drought, leading to extensive desiccation of reservoirs and streams (even the Yellow River flow had repeatedly ceased downstream from Jinan, nearly 200 km from the sea) and to overpumping of aquifers. New Green Wall plantings in arid northwestern provinces were especially affected, above all by the near-record droughts in parts of Shanxi provinces.

Pest problems in China's forests are so severe that the area of annual infestation exceeds the area of total afforestation, and the annual economic loss caused by insect pests equals the state's total investment in forestry (Xinhua, 24 June 1988, *SWB* FE/W0030 A/3). Total annual wood losses caused by pests were put at about 11.5 $Mm^3$, with pine damage by pine moths, extending over some 2.7 Mha, contributing 3.7 $Mm^3$. The other badly affected trees are poplars (with more than 40 per cent of their total stands of 5.3 Mha heavily infested, and losing nearly 6 $Mm^3$ of wood a year) and paulownias (depending on location, 30–80 per cent of these trees are diseased, losing over 1 $Mm^3$ of wood annually.

By 1988 about 4.7 Mha of the Green Wall's shelterbelts (that is, almost half of all plantings) were seriously affected by pests (ranging from pine moths to rats), and about 10 million trees were destroyed (Xinhua, 3 August 1989, *JPRS* CAR 89-005). Perhaps the most worrisome is the massive dying of poplars in the Yanbei section (Shanxi province) of the northern shelterbelt. Plantings of 333,000 ha included 200,000 ha of poplars, but by 1988 31,000 ha of these trees were dead or dying (Xinhua, 29 December 1988, *SWB* FE/W0059 A/4).

Another serious threat to China's tree plantings is the inadequate prevention of forest fires and poor fire-fighting capabilities, failings most dramatically demonstrated by the huge Da Hinggan Mountains fire: between 6 May and 2 June 1987 it destroyed 1.01 Mha of mostly mature, productive coniferous forest, and it also burnt 855,000 $m^3$ of stored timber (Wang 1987).

In total, the north-eastern forest experienced 4,900 fires between 1949 and 1989, burning or damaging about 10 Mha, or roughly eight times the region's newly afforested area during the same period (Fu *et al.* 1989). The best nation-wide estimate is that about a third of successfully established new plantings is eventually damaged by fires.

The combination of false reporting, early sapling mortality caused by poor planting and inadequate care, and pest and fire damage means that, of the 3–5 Mha claimed as newly afforested every year during the late 1980s, at best 1.2–1.8 Mha represent the real net gain. This estimate seems to be confirmed by contrasting two sets of recent Chinese afforestation figures.

Totals provided by the Ministry of Forestry (1990) show China's forest cover declining from 121.86 Mha in the mid-1970s (based on the 1973–76 survey) to 115.28 Mha during the years 1977–81, and then increasing to 124.65 Mha for the period 1984–88. Plantations accounted for 19.4 per cent of the total in the first survey (23.7 Mha), 24.2 per cent in the second (27.8 Mha) and 30.7 per cent in the third (938.3 Mha). These figures document a well-appreciated decline of China's mature forests, but they present afforestation gains averaging about 1.2 Mha a year.

Yet an official announcement claimed that 15.78 Mha were planted between 1986 and 1990, compared to 14.77 Mha during the first half of the 1980s: the total is equivalent to just over 3 Mha a year (Xinhua, 4 March 1991, *SWB* FE/W1030 A/2). The latter figures clearly represent consecutive annual claims forwarded to the centre by local bureaucrats – while the Ministry's surveys come much closer to capturing the reality of long-term survival.

The average rate of about 40 per cent (1.2–1.3 Mha) is better than the 1949–79 record – but the real figure may still be lower. The same Xinhua release put the total area of China's tree plantations at 30.7 Mha in 1990, a figure 20 per cent smaller than the 38.3 Mha total resulting from the latest nation-wide forest survey completed in 1988. Which total is correct?

Interestingly enough, the State Statistical Bureau's annual communiqué on fulfilment of China's national economic plan carried the last nation-wide afforestation claim in its 1984 edition. Since then the communiqués have contained merely general statements about increase in afforested areas, great enthusiasm for tree planting, and new achievements in afforestation, but no specific figures. Nor are these data contained in the most recent editions of the annual *Statistical Yearbook*, although they were always given during the early 1980s, and although the yearbook's content has been generally expanding.

What is most surprising about China's most recent afforestation performance is that the privatized contract management, although undoubtedly improving the average survival rates, has not been able to make a really radical difference. Recent averages around 40 per cent are about 30–40 per cent higher than the typical pre-1980 record, but they still reflect an excessive amount of wasted labour and resources. Of course, even careful planting and the best possible follow-up by rural households specialized in afforestation cannot guarantee high survival rates in environments subject to prolonged droughts and extensive pest and fire damage.

Only higher investment could substantially reduce such risks, but until recently the record has been rather poor. Between 1950 and 1978 the state investment in forestry totalled less than two years of expenditures on water management during the late 1970s – but recent annual capital investment in the state-owned forestry enterprises equalled almost a quarter of that in water conservancy projects (State Statistical Bureau 1990). Also since 1987 the state has been providing discount loans worth 500 million renminbi (Rmb) a year for the establishment of fast-growing tree plantations, and peasants in 16 different provinces are now using a World Bank loan worth US $300 million to set up 1.46 Mha of new forests (Xinhua, 20 January 1991, *SWB* FE/W0164 A/4).

Private investment could not make a major difference as long as the unrealistically low prices of wood precluded a decent rate of return. In 1952 a cubic metre of roundwood sold for Rmb 80, and in 1978 the price was still only Rmb 109; by 1985 the average mixed price (made up of fixed state and negotiated private market prices) was Rmb 313, and it was still difficult to get a good rate of return even with fast-growing species in southern areas. Mixed price per cubic metre rose to almost Rmb 700 by the end of the 1980s: adjusted for inflation this is about 3.5 times higher than in the late 1970s, but given the sharp price increases of all basic production inputs growing wood is still far from a highly profitable proposition.

## Afforestation plans

By the late 1970s, while they were admitting that less than 30 per cent of post-1949 plantings became successfully established, Chinese planners started to talk about forests covering one-quarter of China's territory by the year 2000, an achievement which would have required doubling the existing area to about 250 Mha. As so many other unrealistic plans of the early period of Dengian reforms, these forecasts were soon abandoned.

The latest goals for the 1990s are far from clear. In 1990 the State Council approved a national afforestation plan which called for the planting of 57.16 Mha during the years 1988–2000 (Xinhua, 22 October 1990, *SWB* FE/W0152 A/3). Just a few months later the Minister of Forestry, Gao Dezhan, put the goal at 39 Mha during the 1990s (Xinhua, 12 December 1990, *SWB* Fe/W0159 A/4), and since then some announcements have set the total at 31 Mha. The official claim for the first year of the 1990s was a rather high total of 5.33 Mha (Xinhua, 4 March 1991, *SWB* FE/W1030 A/2).

Long-term plans include extensive afforestation in the middle and upper regions of the Yangzi basin (where some 20 Mha should be replanted by the year 2020), and the establishment of 3.5 Mha of coastal windbreaks by the year 2010 (Xinhua, 15 May 1991, *SWB* FE/W0181 A/3; 30 March 1991, *SWB* FEW0174 A/2).

In order to reduce the pressure on existing forests and to prevent premature harvesting of new plantations, the Ministry of Forestry has decided to limit China's total annual wood consumption (timber and fuelwood) to a maximum of about 245 $Mm^3$ during the 1990s, compared to a harvest of nearly 350 $Mm^3$ during the late 1980s (Xinhua, 21 December 1990, *SWB* FE/W0161 A/6). But it is almost certain that extensive illegal logging will continue and that this limit will be surpassed repeatedly.

## Looking back and looking ahead

There is no doubt that China's post-1949 afforestation efforts represent the world's most sustained and most massive deployment of labour in the history of

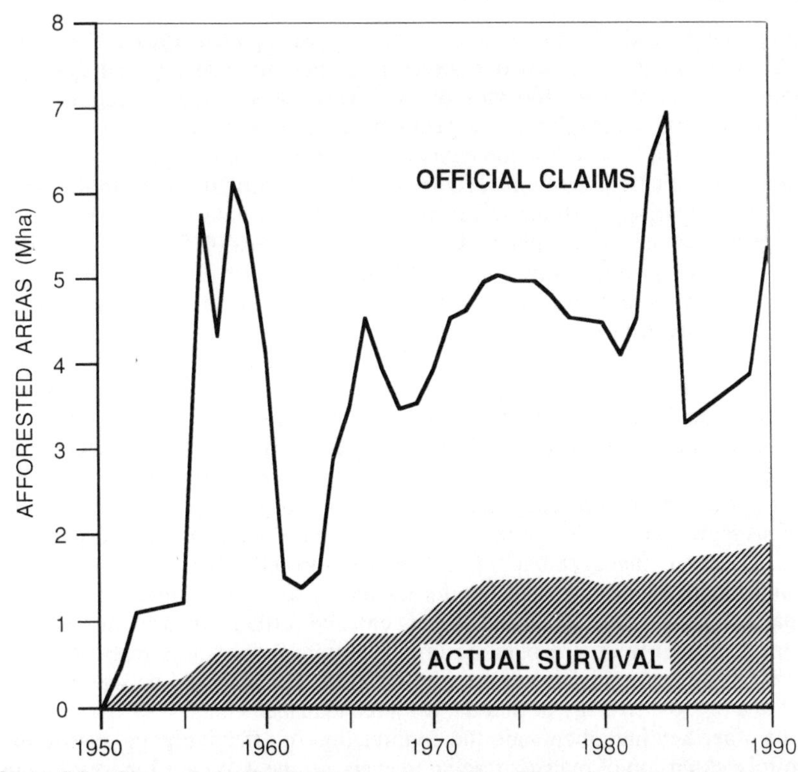


**Figure 8.2** Contrast of official claims and actually afforested land in China between 1950 and 1990. Claimed totals are plotted from data published in *China's Statistical Yearbook* (1980–1984 editions), and from subsequent Xinhua and the Ministry of Forestry news releases. Of course, we do not know what percentage of claimed plantings was successful on an annual basis: area of actual survival is just a suggestive reconstruction, although it adds up to the requisite total of about 40 Mha.

modern forestry, and that, even when making a liberal allowance for totally false claims, tens of millions of hectares were actually planted with new trees. But as with so many other efforts in Communist China, the stress on quantity and the neglect of quality have resulted in an enormous waste of labour and capital. After four decades of prodigious work and claims of more than 130 Mha of afforested land, new plantings have survived on no more than 30–38 Mha, a success rate below 30 per cent (Figure 8.2).

Still, in absolute terms even the lower total looks impressive: 30 Mha is an area about 10 per cent larger than all Swedish forests, and 38 Mha equals the combined area of all forests in France, Germany, Austria, Switzerland and Italy. But such a comparison is grossly misleading: once again, there are enormous qualitative differences involved. China's first forest inventory found that by the late 1970s, 30 Mha of the surviving new plantations, or a quarter of China's forest area, contained less than 200 $Mm^3$ of wood, or merely 2 per cent of all existing timber reserves. Moreover, less than 10 per cent of this small volume was in mature stands ready for commercial harvesting (Zhang *et al.* 1981).

Even if the standing wood reserves in plantations had doubled during the 1980s, the total of about 400 $Mm^3$ of wood is still less than 5 per cent of all late 1980s reserves – although the new plantings now account for about 30 per cent of China's forested area. While the typical growing stock in European forests is 90–100 $m^3$/ha, Chinese plantations average mostly between 10 and 30 $m^3$/ha, and so even in comparison with the lower mean for China's natural stands (70–75 $m^3$/ha), every hectare of new plantings is equivalent to just 0.15–0.40 ha of mature growth. Consequently, today's nearly 40 Mha of Chinese plantations have added an equivalent of perhaps as little as 6 Mha and certainly no more than around 15 Mha of mature growth.

New plantings made an important difference in slowing down the depletion rate of China's mature forests, and in providing timber and fuelwood for local consumption. They have made perhaps their most important ecosystemic contribution as shelterbelts and windbreaks by protecting crops and reducing soil erosion in agricultural areas. But, so far, they have not been able to reverse, or even to stabilize, deforestation and concomitant erosion losses in China's mountainous regions, or to change fundamentally China's wood-supply prospects.

Contractual management of collectively-owned plantings and the re-establishment of private forestry during the 1980s have improved the recent apparent survival rates to around 40 per cent, and further advances should follow as the profitability of tree planting improves with increasingly realistic internal prices. But even highly successful afforestation efforts should be seen only as a part of a broader strategy of rational resource management.

Its other key ingredients are the continuation of effective population-growth controls, extension of realistic pricing to every segment of the Chinese economy, efficient utilization of harvested wood and milling wastes (in general, Chinese use no more than 40–45 per cent of forest phytomass, and only about 7 per cent of processing wastes), and extensive adoption of high-efficiency wood-stoves in rural areas. Without these concurrent steps even a highly effective afforestation programme would be negated in a matter of a few decades.

## References

Ashton, B. *et al.* (1984) *Famine in China*, 1958–61, *Population and Development Review* 10: 613–45.

Chen Tuyan (1983) Large-scale establishment of fuelwood forests is a major avenue to solving the rural energy problem, *Nongye jingji jishu* (Agricultural Economics and Technology) 11: 20–22.

FAO (1982) *Forestry in China*. FAO, Rome.

FAO (1991) *Forest Products Yearbook*. FAO, Rome.

Fu Lixue *et al.* (1989) *Gaishan shengtai huanjing* (*Improving the Environment*). Science Publishing House, Beijing.

Galbraith, J. K. (1973) *China passage*. Houghton Mifflin, Boston.

Li Yongzeng (1989) Chinese forestry: crisis and options, *Liaowang* (Outlook) 12 (20 March): 9–10.

Long Xiangchao (1989) *Jingji cankao* (*Economic Reference*), 15 March 1989, p. 2.

Ministry of Forestry (1981) Run forestry work according to law, *Honggi* (*Red Flag*) 1981(5): 27–31.

Ministry of Forestry (1990) Forest survey data submitted to the World Bank.

Ross, L. (1980) 'Forestry policy in China'. Unpublished PhD dissertation, University of Michigan, Ann Arbor, MI.

Ross, L. (1988) *Environmental policy in China*. Indiana University Press, Bloomington.

Ross, L. and Silk, M.A. (1987) *Environmental law and policy in the People's Republic of China*. Greenwood, New York.

Smil, V. (1984) *The bad earth: environmental degradation in China*. M.E. Sharpe, New York.

Smil, V (1988) *Energy in China's modernisation*. M.E. Sharpe, New York.

Smil, V. (1992) *China's environmental crisis: an inquiry into the limits of national development*. M.E. Sharpe, New York.

State Statistical Bureau (1990) *China's statistical yearbook*. SSB, Beijing.

Wang Jingcai (1978) On scientific afforestation, *Guangming ribao* (*Guangming Daily*), 12 July, p. 4.

Wang Xiaobin (1987) Forest fire scorches bureaucratism, *Beijing Review* 30 (28): 23–25.

Westoby, J. C. (1979) '"Making green the motherland": forestry in China', in N. Maxwell (ed.) *China's road to development*. Pergamon Press, Oxford, p. 243.

Xu Dixin (1981) The position and role of forests in the national economy, *Honggi* (*Red Flag*) 1981 (23):40–45.

Yue Ping (1980) Afforestation and protection of forests is a great cause that brings benefit to the people, *Honggi* (*Red Flag*) 1980(3): 11–16.

Zhang Anghe *et al.* (1981) Better protection and management of forests needed, *Renmin ribao* (*People's Daily*), 25 February 1981, p. 2.

# 9 Forest plantations in Chile: a successful model?

*Antonio Lara and Thomas T. Veblen*

## Introduction

The expansion of forest plantations has often been regarded as an important goal in the forest policy of developing countries, which typically are experiencing rapid deforestation and depletion of their natural forests. Forest plantations have been assumed capable of playing several important roles or functions for society: a) as a basis to assure long-term supply of industrial timber and support industrial expansion; b) to reduce the pressure on natural forests and to improve soil and water conservation; and c) to promote employment and social development in rural areas.

Chile has often been considered a model of successful forest policy because of the expansion and management of forest plantations during the 1970s and 1980s (e.g. Spears 1983; World Resources Institute 1985; Jélvez *et al.* 1990). In addition to the important expansion of timber production and exports based on forest plantations during the 1970s and 1980s in Chile, there also have been important negative social and environmental impacts associated with the expansion of plantations. These impacts should be fully considered in evaluating the overall success of the forest policy based on plantations.

This chapter is concerned with three general questions:

i) Historically, what have been the forest policies that have led Chile to became an important exporter of forest products from plantations?
ii) What are the positive and negative impacts of the forest policy based on the expansion of forest plantations implemented in Chile during the 1970s and 1980s?
iii) What modifications are needed in the forest policy in order to maximize the benefits of forest plantations and correct their negative impacts?

Official forest statistics in Chile include afforestation (i.e. planting in treeless areas) and reforestation under the single category ‘plantations’. These forest plantations consist almost entirely of exotic species. Reforestation includes planting of trees after the final harvest of a plantation as well as the initial conversion of natural forests to human-made forests. Therefore, this chapter refers to forest plantations including: afforestation, reforestation of land previously used for plantations, and reforestation of land previously covered by native forests.

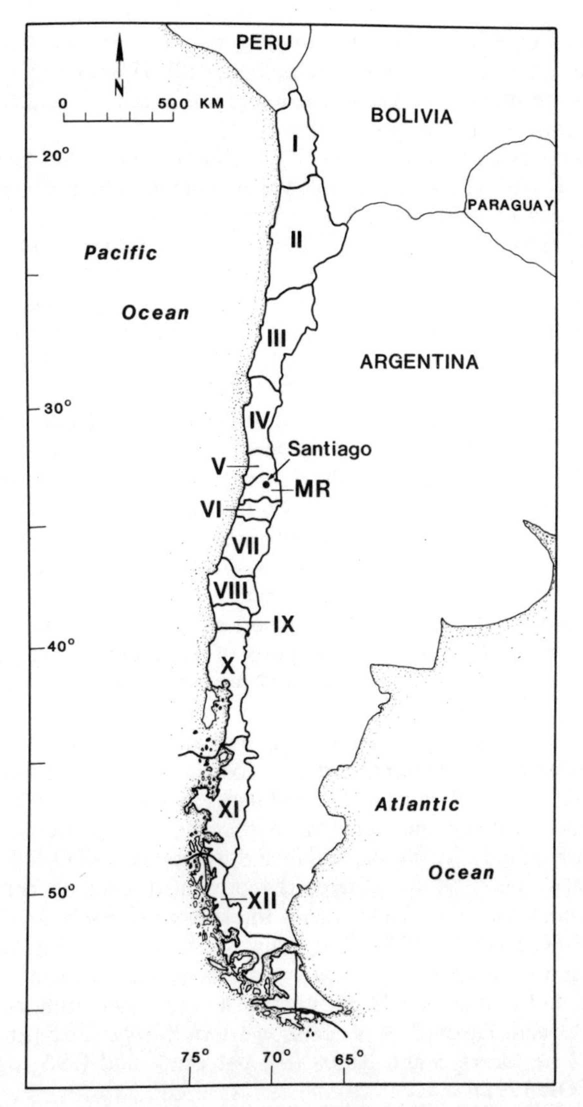


**Figure 9.1** Administrative regions of Chile. MR = Metropolitan Region

## The forestry sector in Chile

Chile has an area of about 755,000 km² (75.5 million ha) and administratively is divided into 13 regions (similar to provinces in other countries) designated I through XII plus a Metropolitan Region in which Santiago, the capital, is located (Figure 9.1). The northern part of Chile (Regions I through IV) has a desert climate and the central part of the country (Regions IV through VIII) has a

Mediterranean-type climate with a southward increase in precipitation. A temperate wet oceanic climate type dominates southern Chile (Regions IX through XII). Native forests are mainly located in Regions IX through XII and forest plantations are concentrated in Regions VII through IX.

In 1990 forest plantations covered 1.45 million ha. There has been no recent national inventory of native forests, but the area covered by potentially productive native forests in 1991 was estimated at 7.61 million ha (CORFO-INFOR 1991). Two main exotic fast-growing species have been planted in Chile: Radiata (Monterey) pine (*Pinus radiata*) and Eucalyptus (*Eucalyptus globulus*), representing 86 per cent and 7 per cent of the total plantation area, respectively (CORFO-INFOR 1991). The total standing timber volume of potentially productive native forests in Chile was estimated in 1991 at 915 million $m^3$, and that of forest plantations at 175 million $m^3$ (CORFO-INFOR 1991).

Resource availability contrasts with industrial forest production. Radiata pine plantations, with about 14 per cent of the forest area in Chile, provide at least 90 per cent of the roundwood for industrial production and exports, excluding fuelwood (CORMA 1991; CORFO-INFOR 1991). This situation does not mean that native forests are being excluded from logging. The main use of timber from native forests is for fuelwood production and the annual harvest for this purpose is estimated at 8–10 million $m^3$ (Lara 1985) compared to 14.2 million $m^3$ harvested for industrial production and export of roundwood and chips in 1990 (CORFO-INFOR 1991). Since 1987, exports of wood chips have increased the pressure on native forests, which provided 52 per cent of the total volume of chips exported in 1990 (CORMA 1991).

Since *c.* 1980 ownership of plantations and native forests used for timber production has been almost entirely private. Today, the state ownership of forests is restricted to *c.* 13.7 million ha of national parks and other protected areas (17.3 per cent of the country), including various cover types (forests, shrublands, grasslands, barren lands, ice sheets, and inland waters) (CORFO-INFOR 1991).

Between 1985 and 1990 forest exports represented about 10 per cent of the total export income of Chile. The value of forest exports reached $855.2 million in 1990 (CORFO-INFOR 1991). The main forest products exported in 1990 were: chemical pulp (37 per cent), sawnwood products (21 per cent), wood chips (13 per cent), and roundwood (9 per cent). The main importing countries and regions in 1990 were Japan (25.6 per cent), western Europe (29.3 per cent), Latin America (15.2 per cent), South Korea (6.4 per cent), and USA (5.9 per cent) (CORFO-INFOR 1991).

## Silviculture and protection of plantations

Radiata pine, introduced in the early 1900s, grows rapidly in the Mediterranean-type and wet oceanic climates of central and south-central Chile. It is usually managed on a 20–25 year rotation with an annual average yield of 20–30 $m^3$/ha. Its wood is mainly used for the production of pulpwood, sawnwood and roundwood for export. Eucalyptus has a similar yield on a 15–25 year rotation. Radiata plantations are commercially thinned for pulpwood starting at age 12 (Jélvez *et al.* 1990). Eucalyptus plantations are managed for timber production and also as coppices for fuelwood production. The main cause of depletion of forest plantations is human-set fires which during the 1990–91 austral summer were estimated to have caused a direct loss estimated at $5.5 million (Chile Forestal 1991a). Large continuous plantations of Radiata pine are particularly susceptible to fire due to the species' flammability and the uninterrupted extent of fuel during the dry summers of the Mediterranean-type climate. During the 1985–90 period an average area of 9,588 ha of plantations was affected by human-set fires each summer (range 2,560–20,349 ha; CORFO-INFOR 1991). In response to the important economic losses due to fire, CONAF (the Chilean Forest Service), in co-operation with large timber companies, has developed a well-organized fire-management programme at an annual cost to CONAF of over $3 million (Chile Forestal 1991a).

## Plantations and forest policy in Chile

### *The pre-1965 period*

Although the aboriginal population of south-central Chile undoubtedly altered native forests through their use of fire (Encina 1954), the commercial exploitation of forests began with Spanish colonization in the sixteenth century. The massive clearing and burning of the forests of the Central Depression and adjacent foothills in the Lake District (Regions IX and X) began in the mid-nineteenth century with the arrival of German colonists. This period of colonization of the Lake District resulted in one of the most massive and rapid deforestations recorded in Latin America until the early 1980s (Veblen 1983).

By the turn of the nineteenth century, extensive areas formerly covered by native forests had been converted to agriculture and pasture land in the areas today included within Regions VI through X (Elizalde 1970; Donoso 1983). A landscape dominated by forests was transformed into one dominated by agriculture and pasture lands. The destruction of native forests and associated soil erosion began to be perceived as important problems as early as 1872 when the first law restricting forest exploitation was passed (Elizalde 1970).

In 1885 Radiata pine was introduced unintentionally and planted along with other North-American conifers as an ornamental species in today's Region VIII, where it grew rapidly (CORMA 1991). Later it became evident that this species produced timber of an adequate quality in planted stands.

Forest destruction continued and in 1931 the first modern forest law was passed. This law included increased and better-organized restrictions on the exploitation of native forests and tax exemptions for a period of 30 years to those lands in which forest plantations were established (CORMA 1991). This was the first law actually promoting afforestation in Chile. At this early stage plantations were promoted for timber production and recovery of deforested areas affected by intense soil erosion.

In 1938 CORFO, a governmental agency for the promotion of primary and industrial production, was created. During the 1940s it stimulated private planting by low-interest loans and partnerships (Haig *et al.* 1946). In 1944 forest plantations covered 143,500 ha, about half located in today's Regions VII and VIII. The

**Table 9.1** Average area planted per year in Chile, 1940–1989 (hectares per year)

| Period | Area planted |
|---|---|
| 1940–44 | 6,091 |
| 1945–49 | 15,825 |
| 1950–54 | 19,910 |
| 1955–59 | 11,958 |
| 1960–64 | 4,392 |
| 1965–69 | 31,570 |
| 1970–74 | 33,814 |
| 1975–79 | 83,033 |
| 1980–84 | 80,683 |
| 1985–89 | 77,425 |

Note: Includes afforestation and reforestation.

*Sources*: 1940–1974, Donoso (1983); 1975–1989, CORFO–INFOR (1991)

two main species that formed those plantations were Radiata pine (58 per cent) and Eucalyptus (31 per cent) (Haig *et al.* 1946). The forestry sector was still mainly based on the exploitation of native forests. Forest industries were still incipient and the main uses of the total annual volume cut were fuelwood (55.4 per cent) and lumber (27.4 per cent). Plantations provided 37 per cent of the volume cut for fuelwood and 18 per cent of the lumber cut (Haig *et al.* 1946).

Incentives given to plantations by the state (tax exemptions, loans, partnerships) promoted the increase of the average annual surface area planted from 1945 through 1954 (Table 9.1). From 1955 through 1964 the average area that was planted significantly decreased due to the temporary lack of interest among private owners.

### *The 1965-73 period*

During this period, Chile was governed by two successive administrations that followed policies of direct participation of the public sector in the economy, redistribution of income, and extensive land reform. The forestry sector was perceived as strategic due to its importance in supplying domestic needs (housing, labour), and also because of the potential for increasing forest exports (Contreras 1989). The government became actively involved in the creation of nurseries, afforestation of public lands, and creation of forest industries, including two major pulpmill complexes (Celulosa Arauco and Celulosa Constitución). In 1972 CONAF was created and became the main governmental agency in the forestry sector (Contreras and Lara 1981; Contreras 1989). In the early 1970s CONAF started a new system of afforestation partnerships known as the 'Convenios de Forestación'. Under this system, small and medium-sized land owners provided the land and CONAF was responsible for establishment, management and logging of the plantation. At the end of the rotation, 25 per cent of the value of the products was expected to belong to the landowner and 75 per cent to CONAF (Contreras 1989). CORFO increased its low-interest loans for affores-

**Table 9.2** Forest products exports, 1960–1990 (millions of US dollars, FOB)

| Year | | Year | |
|---|---|---|---|
| 1960 | 7.7 | 1976 | 166.3 |
| 1961 | 12.1 | 1977 | 206.0 |
| 1962 | 9.8 | 1978 | 255.9 |
| 1963 | 8.4 | 1979 | 403.6 |
| 1964 | 10.2 | 1980 | 580.3 |
| 1965 | 14.5 | 1981 | 415.1 |
| 1966 | 22.5 | 1982 | 337.0 |
| 1967 | 28.7 | 1983 | 323.0 |
| 1968 | 30.4 | 1984 | 372.3 |
| 1969 | 38.2 | 1985 | 317.6 |
| 1970 | 43.5 | 1986 | 393.9 |
| 1971 | 40.2 | 1987 | 558.5 |
| 1972 | 31.7 | 1988 | 719.7 |
| 1973 | 39.1 | 1989 | 783.6 |
| 1974 | 130.9 | 1990 | 855.3 |
| 1975 | 122.6 | | |

*Source*: CORFO–INFOR (1991)

tation on private land as well as its investment in the forest industry (Contreras and Lara 1981; Contreras 1989).

Direct participation of the government in the forestry sector significantly increased in the 1970–73 period; in 1973 CONAF was responsible for 90.4 per cent of the plantations established (INFOR 1980). Forest policy during 1965–73 permitted a significant increase in the rates of afforestation (Table 9.1). Forest production increased at an average of 8.4 per cent per year from 1960 through 1974 (Contreras 1989) and forest exports grew from $10.2 million in 1964 to $43.5 million in 1970 (CORFO-INFOR 1991). They decreased to $39.1 million in 1973 due to the political problems that characterized the 1970-73 period (Table 9.2).

### *The post-1974 period*

The military government that took office after the September 1973 coup began to implement a market economy that stressed the private sector and significantly reduced the direct involvement of the public sector in the economy. Forest policy reflected the view that law enforcement, promotion of the activities undertaken by the private sector and the administration of national parks and other protected areas were the main roles for the government, leaving production in the hands of the private sector.

In 1974 Decree Law 701 (D.L. 701) was promulgated and started to play an important role in the promotion of plantations on private lands (Tables 9.1 and 9.3). D.L. 701 (modified in 1979 by D.L. 2565) provides a 75 per cent subsidy for afforestation costs (later increased to 90 per cent during certain years), as well as

**Table 9.3** Area planted and planting subsidies, 1974–1990

| Year | Area planted (ha) Private sector | Public sector (CONAF) | Total | | Subsidies US $ (Thousands) | US $/ha | Area subsidized/ total area planted the previous year(%) |
|---|---|---|---|---|---|---|---|
| 1974 | 21,052 | 35,171 | 56,223 | | | | |
| 1975 | 38,463 | 44,016 | 82,479 | | | | |
| 1976 | 55,635 | 54,170 | 109,805 | 4,435 | 228 | 0.0 | 11.5 |
| 1977 | 48,499 | 44,637 | 93,136 | 47,174 | 4,617.8 | 97.9 | 84.8 |
| 1978 | 52,486 | 24,885 | 77,371 | 33,674 | 4,282.8 | 127.2 | 69.4 |
| 1979 | 51,749 | 477 | 52,226 | 38,315 | 5,253.8 | 137.1 | 73.0 |
| 1980 | 72,079 | 85 | 72,164 | 45,861 | 6,947.1 | 151.5 | 88.6 |
| 1981 | 92,752 | 29 | 92,781 | 40,502 | 7,675.3 | 189.5 | 56.2 |
| 1982 | 68,545 | 41 | 68,586 | 60,050 | 8,782.9 | 146.3 | 64.7 |
| 1983 | 54,469 | 21,811 | 76,280 | 64,011 | 6,957.8 | 108.7 | 93.4 |
| 1984 | 53,300 | 40,302 | 93,602 | 37,979 | 3,872.2 | 102.0 | 69.7 |
| 1985 | 72,084 | 24,194 | 96,278 | 48,636 | 4,189.2 | 86.1 | 91.2 |
| 1986 | 66,197 | | 66,197 | 49,833 | 5,219.9 | 104.7 | 69.1 |
| 1987 | 65,441 | | 65,441 | 40,947 | 3,992.0 | 97.5 | 61.9 |
| 1988 | 72,508 | | 72,508 | 37,267 | 3,518.4 | 94.4 | 56.9 |
| 1989 | 86,703 | | 86,703 | 30,089 | 2,750.9 | 91.4 | 41.5 |
| 1990 | 94,130 | | 94,130 | 26,695 | 2,910.3 | 109.0 | 30.8 |
| Total | 1,066,092 | 289,818 | 1,355,910 | 605,468 | 71,198.4 | | |

Note: Area planted includes afforestation and reforestation.
*Source*: CORFO–INFOR (1991)

tax exemptions for plantations established from 1974 through 1994 (Ministerio de Agricultura 1974; 1979). The subsidies are paid the year after planting, after the owner has demonstrated a survivorship of at least 75 per cent. D.L. 701 also considers subsidies for pruning and administrative costs. All costs are estimated annually by CONAF. D.L. 701 made reforestation mandatory after the final harvest in all existing and new plantations, at the owner's cost. In addition to the subsidized plantations made by the private sector, CONAF participated actively in the establishment of new plantations from 1974 through 1978 (Table 9.3). In 1974, CONAF established 62.6 per cent of the plantations, 31.6 per cent in 1978 and less than 1 per cent in 1979. Later, CONAF participation was also important from 1983 through 1985 as part of a national programme to reduce unemployment (Table 9.3). The government in practice did not follow its free-enterprise philosophy, as reflected by its direct participation in establishing plantations over several years.

In 1986 CONAF stopped direct planting and transferred this function completely to the private sector. The application of D.L. 701 permitted an increase in the area planted in the 1975–79 period (Table 9.3). The importance of the area planted using the subsidies as compared to the total area planted showed a general increase from 1976 through 1985, and a sustained decrease thereafter (Table 9.3).

This is an indication that D.L. 701 has lessened in importance as a tool to promote plantations by the private sector and that most plantations in 1989 and 1990 were made without the subsidy. The private sector has received $71.1 million in afforestation subsidies that contributed to the establishment of 45 per cent of the area planted in the 1974–90 period (Table 9.3). Subsidies for administrative costs (surveillance, maintenance of fencing and fire-lines) began in 1978 and totalled $12.9 million in the 1978–90 period. In 1986-90 administrative grants for an average of 42,132 ha per year at $4 per hectare were paid. Pruning grants started in 1983 and totalled $6.5 million between 1983 and 1990. In the 1986–90 period pruning grants of an average of 37,915 ha at $26.5 per hectare were paid (from CORFO-INFOR 1991).

Other important ways in which the government has supported the private sector since 1974 are: a) authorization since 1975 to export forest products in any state of processing (including roundwood and chips), which previously was prohibited; b) changes in labour legislation that permitted the use of contract labour and consequently the dramatic reduction of payroll and labour costs for the timber companies; c) transfer of large industrial complexes (e.g. Celulosa Arauco, Celulosa Constitución) as well as land and forests to the private sector at below-market prices, with payment schedules favourable to the private sector; d) planting by CONAF on private lands; and e) support to international marketing efforts and funding of research and technical training (Lara 1985; Contreras 1989). Despite its free-market philosophy the government has favoured the forestry sector compared to other sectors by making heavy use of public money to assist private forestry since 1974. The justification for the special treatment of the forestry sector is based on the assumption of its positive economic externalities.

In March 1990 a democratically elected government took office in Chile and despite the important changes in the political situation of the country, the economic and forest policies have remained little changed.

## Main achievements of the current forest policy

### *Increase in area of plantations, production and exports*

The forest policy implemented since 1974 has been very successful in increasing the area covered by plantations, from about 290,000 ha in 1974 (Jélvez *et al.* 1990) to 1.45 million ha by December 1990 (CORFO-INFOR 1991). Today Chile has the largest Radiata pine holdings in the world (Jélvez *et al.* 1990). Since most of the plantations have not reached rotation age, the total wood volume in them is rapidly increasing. Projections indicate that the annual sustainable harvest from Radiata pine plantations on a sustainable basis will reach 27 million m$^3$ by the year 2000 (INFOR 1987), compared to the 14.2 million m$^3$ harvested in 1990 (CORFO-INFOR 1991).

Another achievement of the forest policy has been a significant increase in production and forestry exports. Production of sawn wood increased by a factor of 3.7 from 1973 to 1990 (0.9 to 3.3 million m$^3$). Chemical pulp production increased from 242,000 tonnes in 1974 to 728,000 tonnes in 1988, but decreased to 644,300 tonnes in 1990. Production of fibre board increased more than six-fold in the 1973-1990 period, from 18,700 $m^3$ to 121,445 $m^3$, and particle board increased more than five-fold in the same period, from 31,230 $m^3$ to 178,292 $m^3$ (CORFO-INFOR 1991). Total exports of forest exports increased from $39.1 million in 1973 to $580 million in 1980 and $855 million in 1990, despite a period of stagnation between 1981 and 1986 (Table 9.2).

*Other achievements*

The current forest policy has also favoured increased investment in forest industries and the improvement of port facilities. Progress in forest legislation has also been important. Under D.L. 701 any logging operation carried out by private owners in plantations or native forests requires a management plan approved by CONAF. Since 1990, logging of plantations may begin as soon as the owner presents a management plan to CONAF. In 1980 Supreme Decree 259 was promulgated, establishing restrictions on the silvicultural systems that could be applied to various forest types and slope classes (Ministerio de Agricultura 1980). Despite the virtues of this decree, logging of native forests based on detrimental silvicultural practices is still a serious problem, due to inadequate law enforcement (Schmidt and Lara 1985; Lara *et al.* 1991).

Another achievement of the current forest policy has been the development of an efficient and well-organized fire-management programme. The current policy has also been important in promoting adequate management of the plantations, subsidizing administrative as well as pruning costs. Soil conservation and land reclamation have been promoted through the grants given for planting grass species (mainly *Ammophila arenaria*) for the control of sand dunes. Subsidized planting of *Atriplex repanda* and *A. nummularia* shrubs in Region IV with the subsidies authorized by D.L. 701 have been important in increasing forage resources and reducing soil erosion in over 35,000 ha of this arid region (CORFO-INFOR 1989).

During the 1970s and 1980s, important progress in the management of many national parks and other protected areas has been made. For example, programmes for the recovery of the populations of some endangered species such as the vicuña (*Vicugna vicugna*), a camelid that lives in the altiplano of Region I, have been successful. CONAF has developed a well-established system of protected areas that despite its limitations may be considered one of the best in Latin America.

Forest research and development have also been promoted through the project 'Investigación y Desarrollo Forestal', funded by CONAF, UNDP (United Nations Development Program), and FAO (Food and Agriculture Organization). Research under this project has mainly been contracted from universities and consultants. In 1991 a project to develop a Forest Action Plan for Chile was initiated. This project is conceived within an international strategy developed by UNDP, World Bank, World Resources Institute and other organizations to promote sustainable forestry in developing countries. The project to develop the Forest Action Plan in Chile is co-ordinated by FAO and is expected to last one year with $1 million in funding (Chile Forestal 1991b).

**Table 9.4** Area planted: Radiata pine and other species, 1978–1990

| Year | Radiata pine | | Other species | |
|---|---|---|---|---|
| | ha | (%) | ha | (%) |
| 1978 | 65,413 | 84.5 | 11,958 | 15.5 |
| 1979 | 48,869 | 93.6 | 3,357 | 6.4 |
| 1980 | 60,086 | 83.3 | 12,078 | 16.7 |
| 1981 | 88,529 | 95.4 | 4,252 | 4.6 |
| 1982 | 61,637 | 89.9 | 6,949 | 10.1 |
| 1983 | 63,884 | 83.7 | 12,396 | 16.3 |
| 1984 | 76,982 | 80.5 | 18,620 | 19.5 |
| 1985 | 80,630 | 83.7 | 15,648 | 16.3 |
| 1986 | 55,058 | 83.2 | 11,137 | 16.8 |
| 1987 | 55,386 | 84.6 | 10,055 | 15.4 |
| 1988 | 61,841 | 84.8 | 11,103 | 15.2 |
| 1989 | 65,587 | 75.6 | 21,118 | 24.4 |
| 1990 | 61,310 | 65.1 | 32,820 | 34.9 |
| Total | 845,212 | 83.1 | 171,491 | 16.9 |

Note: Includes afforestation and reforestation.
*Source*: CORFO–INFOR (1991)

## Main limitations of the current forest policy

### *Resource aspects*

Radiata pine represents 83.1 per cent of the area of plantations established in the 1978–90 period (Table 9.4). Nevertheless, in 1989 and 1990 the dominance of Radiata pine in planting was reduced to 75.7 per cent and 65.1 per cent, respectively, while the area planted with eucalypts increased (Table 9.4). The fact that two exotic species planted as monocultures represent over 93 per cent of the area under plantation in Chile means a restricted diversity of the products that can be obtained. The plantations are also vulnerable to pests and diseases. Native and introduced insects and diseases pose a potential threat to Chile's Radiata pine forests (Ciesla 1988). The European pine shoot moth (*Rhyacionia buoliana*), three species of bark beetle of the family *Scolytidae*, and a foliar pathogenic fungus *Dothistroma pini* are already present in Radiata pine plantations in Chile (Ciesla 1988). Although epidemic damage has not occurred, there is a potential for epidemic attacks that needs to be seriously considered. Aware of this threat, CONAF and the Servicio Agrícola y Ganadero (SAG) have developed a programme of early detection and control of potentially epidemic species in cooperation with the major timber companies.

*Economic aspects*

The philosophy of a free-market economy prevailing in Chile since 1974 rapidly led to a significant economic concentration favouring the development of powerful corporations that control most of the country's economy (Dahse 1979). The forestry sector followed this national pattern and most of the benefits and support given by the government to the forestry sector during the 1970s and 1980s were received by only a few corporations. Recent estimates indicate that four holding companies own about 40 per cent of the forest plantations. Another seven holding companies controlled by foreign capital own 9 per cent of the plantations. The same four holding companies account for almost 70 per cent of the forest exports and along with the foreign corporations are responsible for over 80 per cent of the forest exports (Contreras 1989). A single company (CMPC) produces 90 per cent of the paper in Chile (Contreras 1989). This economic concentration precludes competition and determines oligopsonistic or even regional monopsonistic conditions for the market of land, forests and roundwood, as well as monopolistic or oligopolistic conditions in the domestic market for forest products.

*Labour*

One important disadvantage of the current forest policy is that the dramatic growth in the area of plantations, production and exports has not been coupled with a significant increase in employment in the forestry sector. The average number of full-time jobs in the forestry sector in the years 1966–68 was 63,516. Employment in the forestry sector in 1979–81 averaged 56,921, increasing to 66,476 in 1984 and 82,286 in 1988 (Contreras 1989; CORFO-INFOR 1989). In 1988 it was estimated that silviculture and exploitation provided 41 per cent of the employment in the forestry sector, whereas industry and services provided 44 per cent and 15 per cent respectively (CORFO-INFOR 1989).

The increase of 29.6 per cent in employment in the forestry sector between 1966/1968 and 1988 is much smaller than the dramatic growth in production and exports. In Chile there is an average of one job in silviculture and utilization per 39 ha (from CORFO-INFOR 1989). This is a less intensive use of labour than the one position per 12–20 ha considered adequate for this kind of plantation by international standards (Ohlsson 1976). This difference may be explained by the high proportion of young plantations and by unintensive and often inadequate silvicultural practices. Low salaries, high job instability and inadequate comfort and hygienic conditions in labour camps have also limited the contribution of the forestry sector to the economic and social welfare of the population (Otero 1984).

## Negative impacts of the current forest policy

*Social impacts*

The purchase of land from small, medium-sized and large land owners in rural areas by the timber companies for the establishment of plantations has meant the massive expulsion of the rural population (Rivera and Cruz 1983; Cavieres *et al.* 1986; Otero 1989). Most medium-sized (100–1,000 ha) or large (>1,000 ha) estates sold to establish forest plantations were inhabited by *campesinos* (peasants) on the basis of customary rights with no legal tenure of the land. This population formerly made their living by a combination of small-scale logging of native forests (mainly for fuelwood and charcoal production) and small-scale livestock and agricultural production often within a system of sharecropping with the former land owner (Cavieres and Lara 1983). When the land was purchased by the timber companies, forest dwellers were expelled and forced to migrate. Impoverishment of small land owners (owning estates normally under 50 ha and up to 100 ha) has also led to a generalized process of land transfer to the timber companies and migration (Lara *et al.* 1991). Lack of capital, credit, and technical assistance virtually excluded small owners and significantly limited medium-sized owners from benefiting from the afforestation grants established in 1974. However, afforestation partnerships were very effective for small and medium-sized owners, and 58,400 ha were planted under this system in the 1971–77 period. This system was ended in 1977 and CONAF sold its share to the private sector (Chile Forestal 1980).

A negative relationship between the increase in the area planted and the trend of rural population has been documented for several counties (Cavieres *et al.* 1986; Otero 1989). Migration has been mainly directed to rural villages and small towns (Rivera and Cruz 1983; Cavieres *et al.* 1986; Otero 1989). Poverty, unemployment and lack of services are major problems in these rural villages that have either been created or have significantly increased their population since 1970 (Otero 1989).

### *Environmental impacts*

The expansion of forest plantations in Chile has produced a series of negative environmental impacts. The most dramatic of these has been the destruction of native forests and their conversion to forest plantations. Conversion has been done through clear-cutting and burning of native forests with little or no use being made of the native timber (Lara *et al.* 1989). A total area of 48,600 ha of native forests (mostly second-growth) was destroyed and converted to plantations in Regions VII and VIII in the period between 1978 and 1987. Thirty-one per cent of the native forests of the Coastal Range of Region VIII were converted to plantations in that period (Lara *et al.* 1987). Since D.L. 701 considered only afforestation subsidies, the widespread conversion of native forests to plantations using these subsidies was illegal. The lack of a policy to stop the conversion of native forests and inadequate law enforcement favoured the extensive destruction of native resources (Lara *et al.* 1989).

Forest conversion has been very detrimental to the conservation of three endangered tree species and one endangered shrub species endemic to the Coastal Range of Regions VII and VIII (Lara *et al.* 1989). These species are among the ten woody plant species that are listed as endangered in Chile (CONAF 1985). Forest conversion has also been detrimental to wildlife, including several mammals listed as endangered or vulnerable (CONAF 1988) due to habitat destruction and human-set fires, as well as the use of poisons and herbicides in plantations. Poisons are used to control European rabbit populations that damage young plantations. Herbicides (including Agent Orange) have been used to control introduced woody weeds during the first years of the plantation (Cavieres *et al.* 1986; Lara *et al.* 1989).

Conversion of native forests to plantations has produced an important reduction in biodiversity. In Regions VII and VIII, forest communities with 20 species are being replaced by single-species forests (Lara *et al.* 1991). Remnants of riparian vegetation in the Coastal Range of Region VII that are being extensively replaced by plantations contain 158 vascular plant species, 77 per cent of which are native (San Martín *et al.* 1988).

Conversion has also produced a dramatic reduction of the diversity of the landscape as well as the goods and services from forest lands. Forests with a diversified productivity including timber of different species and qualities, fuelwood, charcoal, edible fruits, forage and shelter for livestock and wildlife, high-quality water, and recreational opportunities are being converted to ecosystems used for the production of a single product: industrial timber (Lara *et al.* 1989).

Clear-cutting and burning during the conversion of native forests to plantations leaves the soil with insufficient cover during the first two or three years of the plantation, causing serious soil erosion during the intense winter rainstorms (Cavieres and Lara 1983). Clear-cutting of mature Radiata pine plantations and slash burning to prepare the land for reforestation is also causing serious soil erosion. Annual soil losses of 35–566 tonnes/ha have been estimated following the logging and burning of Radiata pine plantations in Region VIII. These silvicultural practices are having an important effect in reducing the site productivity during the second rotation of Radiata pine (Alarcón 1992). Burning of about 10,000 ha per year of plantations in Region VIII after logging with an average fuel load of 100 tonnes/ha also makes a significant contribution to atmospheric pollution (Otero 1990).

The expansion of plantations has also had a negative impact on water yield and quality. Radiata pine plantations have higher interception and evaporation rates than those of the native vegetation, which implies decreased water yields in extensively planted watersheds (Huber and Oyarzún 1983). Soil erosion associated with forest conversion and the final harvesting of plantations also produces a significant increase of turbidity in streams (Otero 1990).

Since 1987 the annual area of native forests logged has more than doubled as a result of the use of native roundwood to supply several wood-chip mills. Chips are mainly exported to Japan (Lara *et al.* 1991). Native forests that are being logged to supply the chip-mills (20,872 ha in 1990) are either being converted to Radiata pine and eucalypt plantations or are just being abandoned (Lara *et al.* 1991). Since 1988 there has been a major controversy over the impact of the chipping industry for the conservation of the native forests and the appropriateness of several large projects that are under negotiation with the government (CORMA 1991).

## Recommendations for the improvement of forest policy in Chile

### *Natural forests versus man-made forests?*

An appropriate understanding of the overall goals and performance of the forestry sector in Chile should be the basis for changes in forest policy. The analysis presented thus far indicates the existence of two contrasting forestry sub-sectors in Chile. The first sub-sector is based on forest plantations and provides over 90 per cent of the roundwood for industry and exports. This sub-sector is modern, dynamic, has benefited from subsidies and other forms of support from the government, and has accumulated most of the private investment. The other forestry sub-sector is based on the exploitation of native forests, mainly for fuelwood production, and since 1987 also for wood-chip production for export. Lack of governmental support, inadequate enforcement of protection laws, and exploitation aimed at maximizing the immediate profit without regard for sustainability are leading to the rapid destruction of native forests and their conversion to plantations.

The contrast between forest plantations and native forests has existed since the 1940s but was exacerbated by the forest policy initiated in 1974. Subsidies for plantation forestry, in contrast to lack of government support for the management and recuperation of native forests, have created a distortion in the decision-making process and have promoted the conversion of native forests.

The bleak situation of the native forests contrasts with their high potential under sustainable management practices. Research done at several universities has estimated annual yields of 9–14 $m^3$/ha for thinned second-growth forests, compared to 20-30 $m^3$/ha for Radiata pine and Eucalyptus plantations (Schmidt and Lara 1985). However, prices of native timber are normally between two and four times higher than those of Radiata pine. Therefore, the management of second-growth native forests under sustained-yield practices is economically feasible and may be at least as profitable as the management of forest plantations (Cavieres and Lara 1983; Lara 1985).

### *The need for a balance between development based on plantations and on native forests*

The disparity between availability of resources promoting plantation forestry and those encouraging the management of native forests, plus the high economic potential of the latter, suggests the need for major changes in the current forest policy. There should be a shift towards a balanced development of both forestry sub-sectors. Such a policy should acknowledge the interdependence and distinctive characteristics of each sub-sector and should end the process of conversion of native forests. The natural differentiation between plantations and native forests in terms of geographic distribution as well as products and services that each can provide suggests the need to replace the historically antagonistic relationship between these sub-sectors with a complementary one. The need for afforestation in important areas deforested in the last century and now suffering from high erosion rates (Peralta 1978; Veblen 1983) also illustrates the inadequacy of the current forest policy, which has resulted in massive conversion of native forests.

**Table 9.5** Geographical distribution of plantations, 1974–1990

| | REGION | | | | | | | | | | | | | |
|---|---|---|---|---|---|---|---|---|---|---|---|---|---|---|
| | I | II | III | IV | V | M.R. | VI | VII | VIII | IX | X | XI | XII | TOTAL |
| Year | | | | | | | | | | | | | | |
| 1974 | | | | 283 | 1,876 | 535 | 3,102 | 9,165 | 28,555 | 6,435 | 3,012 | 3,220 | | 56,183 |
| 1975 | | | | 31 | 1,555 | 1,121 | 2,424 | 10,292 | 44,123 | 11,115 | 7,613 | 4,205 | | 82,479 |
| 1976 | 30 | | | 756 | 2,061 | 2,963 | 12,650 | 10,885 | 45,105 | 17,950 | 13,544 | 1,861 | | 107,805 |
| 1977 | | | | 1,609 | 2,473 | 2,961 | 7,565 | 11,119 | 40,424 | 13,358 | 12,353 | 1,310 | | 93,172 |
| 1978 | | | | 3,300 | 3,789 | 1,245 | 10,691 | 9,192 | 32,327 | 8,794 | 6,686 | 1,336 | 11 | 77,371 |
| 1979 | | | | 1,019 | 694 | 341 | 1,611 | 6,942 | 28,457 | 6,436 | 5,245 | 1,481 | | 52,226 |
| 1980 | 50 | | | 6,259 | 1,837 | 1,804 | 2,344 | 13,522 | 31,306 | 9,959 | 4,176 | 907 | | 72,164 |
| 1981 | | | | 1,998 | 1,521 | 176 | 2,426 | 21,616 | 41,546 | 16,393 | 6,743 | 362 | | 92,781 |
| 1982 | | | 25 | 3,906 | 480 | 83 | 1,285 | 13,371 | 32,754 | 13,170 | 3,085 | 427 | | 68,586 |
| 1983 | 181 | | 165 | 5,295 | 1,423 | 669 | 3,721 | 14,332 | 30,653 | 14,786 | 4,234 | 821 | | 76,280 |
| 1984 | 325 | | 82 | 5,579 | 2,337 | 1,415 | 7,005 | 22,446 | 26,824 | 17,987 | 8,370 | 1,232 | | 93,602 |
| 1985 | 391 | 37 | 101 | 3,949 | 2,634 | 726 | 6,856 | 22,783 | 30,207 | 18,890 | 8,642 | 1,021 | 40 | 96,277 |
| 1986 | 38 | 15 | | 2,438 | 1,898 | 122 | 2,235 | 15,971 | 20,864 | 14,885 | 7,048 | 583 | | 66,097 |
| 1987 | 5 | 3 | 6 | 2,773 | 1,502 | 65 | 1,507 | 14,043 | 23,729 | 13,286 | 7,490 | 1,032 | | 65,441 |
| 1988 | 20 | 9 | 8 | 626 | 2,359 | 75 | 2,076 | 15,382 | 30,445 | 12,522 | 8,724 | 262 | | 72,508 |
| 1989 | | | 13 | 1,929 | 3,466 | 64 | 4,519 | 16,223 | 35,329 | 15,812 | 8,986 | 362 | | 86,703 |
| 1990 | | | 125 | 841 | 5,828 | 151 | 4,311 | 13,944 | 44,795 | 10,495 | 13,183 | 457 | | 94,130 |
| Total | | | | | | | | | | | | | | |
| ha | 1040 | 64 | 525 | 42,591 | 37,733 | 14,516 | 76,328 | 241,228 | 567,443 | 222,273 | 129,134 | 20,879 | 51 | 1,353,805 |
| % | 0.1 | 0.0 | 0.0 | 3.1 | 2.8 | 1.1 | 5.6 | 17.8 | 41.9 | 16.4 | 9.5 | 1.5 | 0.0 | 100.0 |

Notes: Includes afforestation and reforestation. There are small differences in the totals compared with Table 9.3.
*Source: CORFO*–INFOR (1991)

*The need for species diversification in forest plantations*

Claims for the need to increase the diversity of plantations have been made since 1942 (Donoso 1983). The great environmental heterogeneity of Chile and the different roles that forest plantations may play in the various regions, indicate a need for plantation of species other than Radiata pine and eucalypts. Successful experimental or small-scale plantations have indicated the suitability of several native and exotic species for afforestation projects. For example, plantations of *Prosopis tamarugo* (a native species of mesquite) have been established since the 1950s in Region I for forage and fuelwood production. These plantations have expanded the remnant native *P. tamarugo* forests that grow in desert areas as phreatophytes. In Region XI thousands of hectares of native forests were destroyed by human-set fires in the 1940s, triggering severe soil erosion (Peralta 1979; Veblen 1983). In this region several exotic species, including Ponderosa pine (*Pinus ponderosa*), Lodgepole pine (*P. contorta*), Douglas fir (*Pseudotsuga menziesii*) and others, have grown well. Since 1974 more than 20,000 ha have been afforested with these species (Table 9.5).

The high potential of some native Chilean species in forest plantations has been demonstrated both in Chile and in other countries. The practices for the reproduction of most of the native tree species have been investigated, and today native trees of different species are produced in several nurseries in Chile. Experimental plantations established in 1959 in Region X demonstrated that various species of southern beech (*Nothofagus obliqua*, *N. alpina* (synonym *N. procera*) and *N. dombeyi*) have yielded promising results (Vita 1977). Experimental plantations with *N. pumilio* in Region XI have also been initiated (Schlegel *et al.* 1979). Preliminary data indicate potential annual yields of 20 $m^3$/ha or higher for *N. dombeyi* plantations and 8–18 $m^3$/ha for *N. alpina* plantations (Donoso *et al.* 1991a; 1991b). *N. obliqua* and *N. alpina* stands have been planted in several locations in England, Scotland and Wales since 1930. These plantations have demonstrated an annual yield of 12–14 $m^3$/ha and economic returns substantially greater at all rates of discount than those likely from oak or native beeches (Nimmo 1971). *N. alpina* and *N. obliqua* have been also planted in Denmark, Germany and France. The concern over air-pollution damage to conifers in Europe has recently prompted an increased interest in the plantation of *Nothofagus* species (Destremau 1988).

The examples already discussed indicate the feasibility of a forest policy directed towards the diversification of plantations, using exotic and native species other than Radiata pine and eucalypts. The change from mono-specific plantations towards mixed plantations has also been recommended. Mixed plantations of Radiata pine and *Acacia melanoxylon*, for example, have been proposed for Region VIII because of the potential role in atmospheric nitrogen fixation, rapid growth and high price for the latter (Otero 1990).

*The need for a better geographic distribution of plantations*

About 60 per cent of the plantations established in the 1974–90 period are in Regions VII and VIII, and 86 per cent are located in Regions VII through X

(Table 9.5). An increase of plantations in other regions would be beneficial for the increase and conservation of forest and soil resources as well as for economic and social development. A more even geographic distribution of plantations would also reduce the pressure towards the conversion of native forests and the plantation on soils suitable for pastures such as has occurred in Regions VIII through X.

### *The need of diversification of products and services from plantations*

The diversification of plantations towards the production of goods and services other than industrial timber should also be a priority. Adequately planned and managed plantations can play an important role in fuelwood production, soil and water conservation as well as forage production. Radiata pine plantations with lower planting densities than those used traditionally have demonstrated a good potential for a combined use of livestock and timber (Sotomayor 1991). Available data indicate that fuelwood and charcoal supply over 20 per cent of the total energy consumed in Chile (CNE 1982) and that most of the fuelwood comes from native forests (Lara 1985). Plantations may become important for fuelwood production with an important social effect as well as a reduced pressure on the native forests.

### *The key role of subsidies and other incentives in forest policy*

Governmental incentives such as afforestation grants and partnerships have been powerful tools in forest policy in Chile. The experience also indicates how subsidies in the forestry sector may have undesirable social and environmental impacts.

One important tool to implement the changes that have been proposed here would be through the modification of the subsidies in the forestry sector. Governmental support to the forestry sector needs to be reoriented towards the social groups and productive activities that did not receive enough attention in the 1970s and 1980s. Management of native forests, diversification of forest plantations and promotion of rural development should be targeted as priorities for subsidies and other governmental incentives. The increasing ability of the timber companies to maintain high planting rates in industrial plantations with a decreasing dependence on planting grants (Table 9.3) demonstrates the feasibility of reducing afforestation subsidies to the large timber companies.

The balanced development of the forestry sector based on plantations and native forests requires governmental support for both. The need for subsidies for the management of native forests was proposed by the Association of Foresters in 1979 (Cavieres and Lara 1983) and CONAF prepared a first draft of such a policy in 1984 (CONAF 1984). Recently, CONAF prepared a project for a law that would subsidize some silvicultural practices in native forests (CONAF 1991), but its success is still uncertain. A significant improvement of law enforcement, promotion of research, technology transfer and marketing are also essential for promoting the sustainable management of native forests.

Subsidies and other governmental incentives may also play a key role in promoting species diversification in plantations, a more even geographical distribution and diversification of goods and services from plantations. Specific subsidies should be established to promote the planting of native species, planting in geographical regions with a small proportion of forest plantations, and for uses that include fuelwood production, soil and water conservation, and range production.

In framing conditions for afforestation subsidies, consideration should be given to limitations on the maximum size of continuous stands, as well as to the proportion of watersheds that can be covered by plantations. Recipients of subsidies should be subject to regulations about harvesting methods, the use of fire to reduce slash, and other silvicultural practices. These regulations, combined with adequate enforcement, would reduce current soil and water conservation problems as well as the vulnerability of plantations to damage by fire and pests.

Since the subsidies have not been effective in reaching small and medium-sized land owners, other incentives should be considered. Partnerships, technical assistance, and low-interest loans may play an important role in promoting rural development and forest conservation among these land owners.

Adequate administration of the governmental incentives would require an increase in the resources available to CONAF. Law enforcement is the necessary complement to the government incentives to assure an adequate management of the forest resources. Co-ordination with other policies (e.g. housing, energy) is also needed to improve the contribution of the forestry sector to the social needs of the country.

## Conclusions

Forest plantations have played an important and increasing role in the development of the forestry sector in Chile. Two exotic species (Radiata pine and Eucalyptus) have been successfully planted and managed since the 1920s, and became important in supplying roundwood for the industry in the 1940s. Public-sector support has been a key factor in the expansion of forest plantations and industry since 1964. In the period from 1964 to 1973, economic and forest policy based on the direct participation of the public sector permitted a significant increase of plantations and expansion of forest industries. Since 1974 economic and forest policy based on a free-market economic philosophy has ironically provided strong governmental support to the large timber companies by way of afforestation subsidies, tax exemptions, transfer of industries, forest and lands, and labour laws favourable to industry. The private sector also benefited from a law that authorized the export of roundwood and chips. This policy has been effective in the expansion of forest plantations and the significant growth of production and exports. The forest policy implemented since 1974 has also been fundamental in making forestry a profitable business for the large timber companies. Resources developed during several decades of state support (i.e. industries, plantations, experience, technical knowledge, etc.) have been an important basis for the growth of the forestry sector since 1974. Projections indicate that the annual sustainable harvest from plantations could be doubled by the year 2000, when many of today's young plantations reach rotation age.

In contrast to these achievements, forest policy implemented since 1974 has been limited in its capacity to increase employment and welfare among forest workers. This policy, together with the country's economic policy, has produced strong economic concentration in the forestry sector, precluding competition and a more even distribution of resources. The current forest policy has produced negative social impacts, especially the massive expulsion of forest-dependent people (mainly *campesinos*) who have migrated to villages and small towns.

The forest policy implemented in Chile since 1974 has also produced a series of negative environmental impacts such as: massive conversion of native forests that include some threatened flora and fauna species; reduction of biodiversity; increase of soil erosion, and decrease of water yield and quality.

Significant changes in Chilean forest policy are recommended. These changes refer to a balanced development of both forest plantations and native forests, diversification of plantations in terms of species composition, geographical distribution and objectives of the plantations (i.e. combinations of timber production, fuelwood, forage, soil and water conservation, etc.). Planting grants, afforestation partnerships, low-interest loans, and technical assistance, along with improved law enforcement, are important mechanisms for a balanced development of the forestry sector in Chile. Governmental support for small and medium-sized land owners should be a priority.

Experience during the 1970s and 1980s demonstrates that governmental action in collaboration with the private sector can have a significant effect on the expansion of forest plantations in developing countries. This has created the basis for providing long-term supply of industrial timber and has significantly increased industrial production and exports. At the same time, however, experience shows how some negative social and environmental impacts may significantly limit or diminish the overall social and economic value of plantation forestry. Experience indicates the need to consider explicit mechanisms to promote social development and resource conservation in conjunction with forest-plantation policy. These mechanisms cannot be assumed to be the automatic consequences of the expansion of industrial forest plantations. The potential of forest plantations to reduce the pressure on native forests, improve soil and water conservation and promote employment and social development has not been realized in Chile and constitutes a major future challenge.

## References

Alarcón, C. (1992). *Evaluación económica de los efectos de pérdida de productividad de sitio en la segunda rotación.* Instituto Forestal, Santiago.

Cavieres, A. and Lara, A. (1983) *La destrucción del bosque nativo para ser reemplazado por plantaciones de pino insigne: Evaluación y proposiciones. I. Estudio de caso en la Provincia de Bíobío.* CODEFF Informe Técnico No. 1. Santiago.

Cavieres, A., Martner, G., Molina, G. and Paeile, V. (1986) Especialización productiva, medio ambiente y migraciones. El caso del sector forestal chileno, *Agricultura y Sociedad* 4/86: 31–95.

Chile Forestal (1980) Plantaciones forestales, *Suplemento Chile Forestal* 54.

Chile Forestal (1991a) Temporada de incendios 90–91, *Chile Forestal* 184: 25–26.
Chile Forestal (1991b) Un Plan de Acción Forestal para Chile, *Chile Forestal* 183: 8.
Ciesla, W.M. (1988) Pine bark beetles: a new pest management challenge for foresters, *Journal of Forestry* 86: 27–31.
CNE (1982) *Balance nacional de energía 1963–1982*. Comisión Nacional de Energía, Santiago.
CONAF (1984) *Ideas preliminares para un proyecto de bonificación por manejo de renovales y por enriquecimiento del bosque nativo*. CONAF Gerencia Técnica Departamento de Control Forestal, Santiago.
CONAF (1985) *Simposio flora nativa arbórea y arbustiva de Chile amenazada de extinción*. Santiago.
CONAF (1988) *Simposio fauna nativa de Chile amenazada de extinción*. Santiago.
CONAF (1991) *Proyecto de ley sobre recuperación del bosque nativo y fomento forestal.* Santiago.
Contreras, R. (1989) *Más allá del bosque. La explotación forestal en Chile*. Editorial Amerindia, Santiago.
Contreras, R. and Lara, A. (1981) El traspaso de funciones del estado en el sector forestal, *Boletín Estudios Agrarios GEA* 7: 1–24.
CORFO-INFOR (1989) Estadísticas forestales 1988, *Boletín Estadístico* No. 11. Corporación de Fomento de la Producción/Instituto Forestal, Santiago.
CORFO-INFOR (1991) Estadísticas forestales 1990, *Boletín Estadístico* No. 21. Instituto Forestal filial CORFO, Santiago.
CORMA (1991) *Chile, país forestal*. Departamento del Bosque Nativo, Corporación Chilena de la Madera, Santiago.
Dahse, F. (1979) *Mapa de la extrema riqueza. Los grupos económicos y el proceso de concentración de capitales*. Editorial Aconcagua, Santiago.
Destremau, D.X. (1988) 'La sylviculture des Nothofagus en Europe', in *Simposio sobre Nothofagus: Monografías de la Academia Nacional de Ciencias Exactas Físicas y Naturales*, Buenos Aires, pp 115–20.
Donoso, C. 1983. 'Modificaciones del paisaje forestal chileno a lo largo de la historia', in *Proceedings of the Symposium 'Desarrollo y perspectivas de las disciplinas forestales de la Universidad Austral de Chile'*, Valdivia, pp 365–438.
Donoso, C., Escobar, B. and Cortés, M. (1991a) *Técnicas de vivero y plantación para Raulí (Nothofagus alpina)*. Documento Técnico 53 Chile Forestal, Santiago.
Donoso, C., Escobar, B. and Cortés, M. (1991b) *Técnicas de vivero y plantaciones para Coigüe (Nothofagus dombeyi)*. Documento Técnico 55 Chile Forestal, Santiago.
Elizalde, R. (1983) *La sobrevivencia de Chile*. Ministerio de Agricultura, Servicio Agricola y Ganadero, Santiago.
Encina, F.A. (1954) *Resumen de la historia de Chile*, Vol 1. Editorial Zig-Zag, Santiago.
Haig, I.T., Teesdale, L.V., Briegleb, P.A., Payne, B.H. and Haertel, M.H. (1946) *Forest resources of Chile as a basis for industrial expansion*. Forest Service, US Department of Agriculture, Washington, DC.
Huber, A. and Oyarzún, C. (1983) Precipitación neta e intercepción en un bosque adulto de *Pinus radiata* D. Don, Bosque 5: 13–20.
INFOR (1980) *Estadísticas Forestales 1979*. Instituto Forestal, Santiago.
INFOR (1987) *Disponibilidad de madera de pino radiata en Chile 1986–2015*. Instituto Forestal Informe Técnico 103, Santiago.
Jélvez, A., Blatner, K.A. and Govett, R.L. (1990) Forest management and production in

Chile, *Journal of Forestry* 88: 30–34.

Lara, A. (1985) Los ecosistemas forestales en el desarrollo de Chile, *Ambiente y Desarrollo* 1(3): 81–99.

Lara, A., Araya, L., Capella, J. and Fierro, M. (1987) *Evaluation of the destruction of native forests in south-central Chile.* Final Report Project #3181. WWF International/ CODEFF, Santiago.

Lara, A., Araya, L., Capella, J., Fierro, M. and Cavieres, A. (1989) *Evaluación de la destrucción y disponibilidad de los recursos forestales nativos en la VII y VIII Región.* CODEFF Informe Técnico, Santiago.

Lara, A., Donoso, P., and Cortés, M. (1991) *Development of conservation and management alternatives for native forests in south-central Chile.* Final Report Project # 3181. WWF-US/ CODEFF, Santiago.

Ministerio de Agricultura (1974) *Decreto Ley 701 de Fomento Forestal.* Santiago.

Ministerio de Agricultura (1979) *Decreto Ley 2565. Nuevo Decreto Ley 701 de Fomento Forestal.* Santiago.

Ministerio de Agricultura (1980) *Decreto Supremo 259. Reglamento del Decreto Ley 701 sobre fomento Forestal.* Santiago.

Nimmo, M. (1971) *Nothogfagus plantations in Great Britain.* Forestry Commission Forest Record 79.

Ohlsson, B. (1976) 'La actividad forestal como fuente de empleo', in *FAO/SIDA Seminario sobre ocupación forestal en América Latina*, Lima, pp 15–36.

Otero, L. (1984) 'Caracterización laboral, estudio de las condiciones de trabajo y análisis ocupacional de los trabajadores forestales en la Octava Región del país'. Thesis, Facultad de Ciencias Agrarias, Veterinarias y Forestales, Universidad de Chile, Santiago.

Otero, L. (1989) 'Aldeas rurales y plantaciones forestales: un estudio sobre la relación entre la población y su medio natural'. Thesis, Facultad de Arquitectura y Bellas Artes, Instituto de Estudios Urbanos, Pontificia Universidad Católica de Chile, Santiago.

Otero, L. (1990) Impacto de la actividad forestal en comunidades locales en la VIII Region, *Ambiente y Desarrollo* 6(2): 61–69.

Peralta, M. (1978) Procesos y areas de desertificación en Chile continental: mapa preliminar, *Ciencias Forestales* 1(1): 41–44.

Peralta, M. (1979) 'Los procesos erosivos y la erosión', in IREN-CORFO (eds.) *Perspectivas de desarrollo de la Región de Aysén del General Carlos Ibánez del Campo. Suelos y Erosión.* Santiago, pp 83–89.

Rivera, R. and Cruz, M.E. (1983) La realidad forestal chilena. *GIA Resultados de Investigaciones No 15.* Santiago.

San Martín, J., Troncoso, A. and Ramírez, C. (1988) Estudio fitosociológico de los bosques pantanosos nativos de la Cordillera de la Costa en Chile Central, *Bosque* 9(1): 17–33.

Schlegel, F., Veblen, T. and Escobar, B. (1979) Estudio ecológico de la estructura, composición, semillación y regeneración del bosque de Lenga (*Nothofagus pumilio*) XI Región. *Informe de Convenio No 8. Serie Técnica*, Facultad de Ingeniería Forestal, Universidad Austral de Chile, Valdivia.

Schmidt, H. and Lara, A. (1985) Potencialidad de los bosques nativos chilenos, *Ambiente y Desarrollo* 1(2): 91–108.

Sotomayor, A. (1991) *Sistemas silvopastorales y su manejo (II Parte)*, Documento Técnico 43 Chile Forestal, Santiago.

Spears, J.S. (1983) Tropical reforestation: an achievable goal? *Commonwealth Forestry Review* 62: 201–17.

Veblen, T.T. (1983) 'Degradation of native forest resources in southern Chile', in H.K. Steen (ed.) *History of sustained-yield forestry: a symposium.* Forest History Society, Durham, North Carolina, pp 344–52.

Vita, A. (1977) Crecimiento de algunas especies forestales nativas y exóticas en el arboretum del Centro Experimental Forestal, *Boletín Técnico* no. 47, Facultad de Ciencias Forestales, Universidad de Chile, Santiago.

World Resources Institute (1985) *Tropical forest action plan: a call for action.* World Resources Institute, Washington, DC.

# 10 New Zealand: afforestation policy in eras of state regulation and deregulation

*M.M. Roche and R.B. Le Heron*

## Introduction

For many New Zealanders, 'forestry' is synonymous with afforestation and with exotic tree species of which the principal one is the Monterey pine (*Pinus radiata*), a species maturing in New Zealand in 25 to 30 years and suitable for pulping at even younger ages. By 1983 a century's endeavour, including two major eras of afforestation (the first in the 1920s and 1930s, the second in the 1960s, 1970s and 1980s), culminated in the millionth hectare of forest being established in New Zealand.

Afforestation in New Zealand has been influenced by the strategic and commercial opportunities conferred by rapidly growing exotics. These trees allowed the state in the 1930s, against a backdrop of extensive pioneer clearing of the natural forests, to step beyond orthodox forestry principles stressing management of natural forests, and to embark on large-scale afforestation. The timespan for maturity was just short enough for forestry to have private-sector appeal as a long-term investment. In consequence, large areas of the central North Island and to a lesser extent most of the other districts of New Zealand have some state, company or other private exotic forest plantations.[1]

Industrial forestry in New Zealand depends on exotic plantation forests. The increased quantities of wood that will become available from these forests will further integrate New Zealand into the global industrial forestry scene, especially as the local market stabilises. While presently distinctive, the New Zealand experience may be a precursor of further integrated developments in forestry as other nations seek to expand industrial plantation forestry.

This chapter examines New Zealand's afforestation experience, with emphasis on developments since 1980. The chapter is concerned less with describing the observable pattern of forest areas and species types, and more with placing afforestation activity in a wider regional, national, and international forestry context. The latter is especially important since the globalisation of the New Zealand forestry sector has been one of the major outcomes of the government's deregulatory thrust in the mid-1980s. Afforestation is discussed as a land use arising in a capitalist context, from the investment activity of both private- and public-sector actors. The restructuring of state forestry from 1984 to 1988 is also examined and some of the implications of the resulting privatisation of exotic plantation forests are considered. The focus then shifts to implications of the

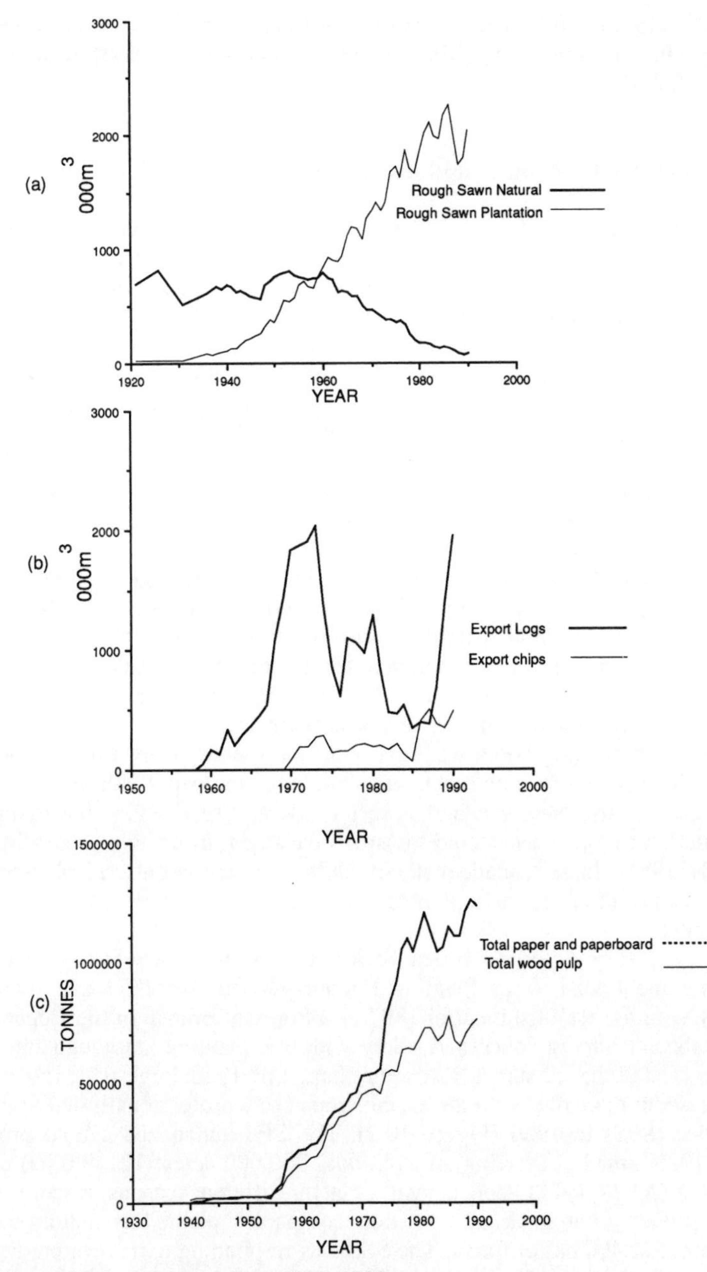


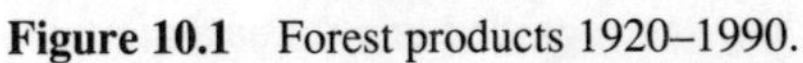

**Figure 10.1** Forest products 1920–1990.
(a) Wood production from natural and plantation forests.
(b) Chip and log exports from plantation forests.
(c) Production of wood pulp and paper and paperboard.

*Source*: Ministry of Forestry (1991a)

integration of plantation forests into globally based forestry networks. Finally, the efforts of the Ministry of Forestry (MoF) in 1991 to increase annual afforestation rates are scrutinised.

## New Zealand afforestation in a capitalist context

In order to understand New Zealand's afforestation experience, we must move beyond a collation of historically specific details concerning areas planted, species chosen, land-use change and conflict, and locational patterns of processing facilities. It is important to recognise that afforestation in New Zealand needs ultimately to be discussed as a capitalist investment process. A central underpinning of most afforestation activity in New Zealand has been the creation of value and the transformation of natural resources by production systems organised to this end. Initially this meant the exploitation of the indigenous forests of New Zealand, already reduced by natural fires and Maori burning to approximately half their earlier extent, to cover 50 per cent of the land area at the beginning of widespread European settlement in 1840. The indigenous forests were further diminished to 23 per cent of the surface area by 1900. Most reduction arose from land settlement, rather than from the activities of the timber industry. Indigenous forest species (especially Rimu (*Dacrydium cupressinum*)) continued to be the mainstays of the domestic timber markets until the 1950s (Figure 10.1). Indeed, not until 1960 did national timber production from exotic plantations exceed that from the indigenous forests. This point was reached in 1955 in the North Island and 1965 in the South Island (Ministry of Forestry 1991a).

The 1920s state planting boom was first seen as a source of sawn timber, but by the early 1930s state planning had begun for a pulp and paper industry in the 1940s. Simultaneously, New Zealand Forest Products Ltd (NZFP), the major private plantation owner, was steered towards investment in its own processing plants (Healy 1981). In fact, inadequate silvicultural treatment at critical times meant that much of the forest area planted in the 1920s was of a quality suitable only for pulping.

Private afforestation experiments date back to the 1850s, and the state became involved with the Forest Trees Planting Encouragement Act 1871 and more substantially with the establishment in 1897 of a Forestry Branch of the Department of Lands and Survey concerned solely with tree planting. Although some 37,100 acres (15,000 ha) of state forest were planted by 1920, large-scale afforestation came about thereafter with the establishment of a professionally-led State Forest Service (SFS) in 1920 (Figure 10.2). The SFS announced a bold programme in 1925 aimed at creating an additional 300,000 acres (121,400 ha) of forest by 1935 (*AJHR* 1925). Concurrently, but for different reasons, a score of public and private joint-stock afforestation companies planted an additional 300,000 acres (121,400 ha) of forest. The SFS was responding to its own prediction of a national timber famine by the 1970s, and regarded exotics as an alternative timber resource, giving scope to regenerate indigenous forests and introduce sustained yield management. For the afforestation companies the situation was less complex: a timber famine represented a marvellous commercial opportunity to create a plantation estate which was likely to offer good returns 25 years hence

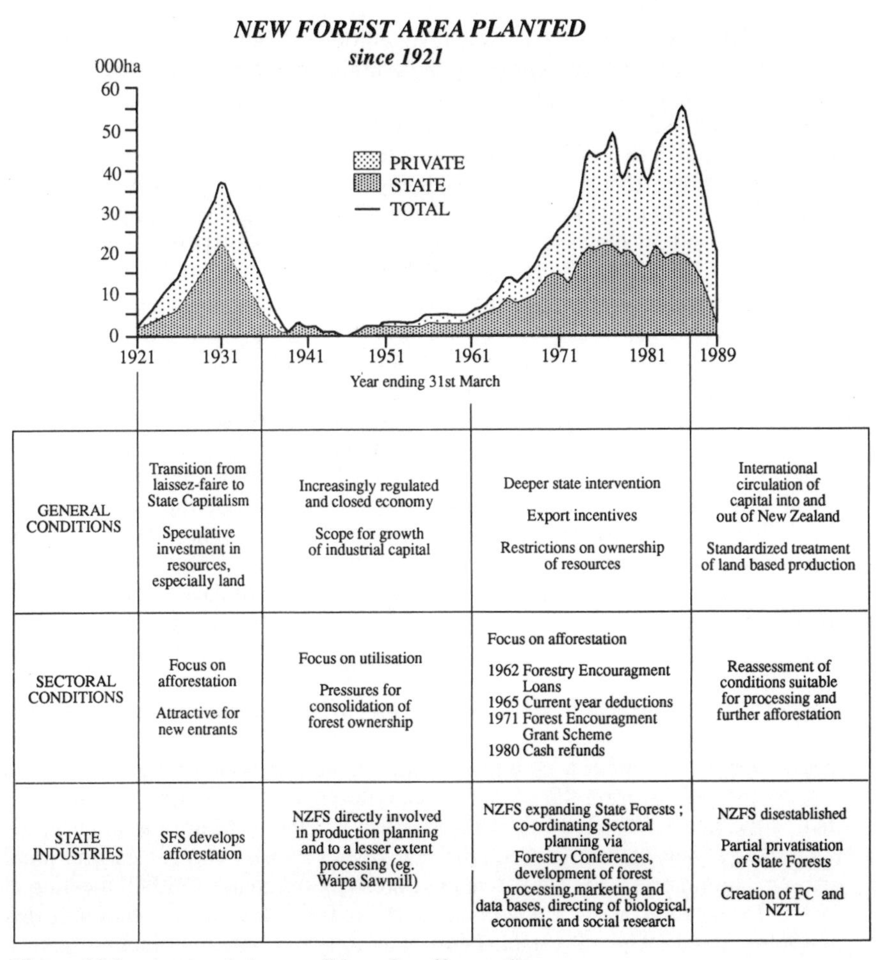


| | 1921–1941 | 1941–1961 | 1961–1983 | 1983–1989 |
|---|---|---|---|---|
| GENERAL CONDITIONS | Transition from laissez-faire to State Capitalism<br>Speculative investment in resources, especially land | Increasingly regulated and closed economy<br>Scope for growth of industrial capital | Deeper state intervention<br>Export incentives<br>Restrictions on ownership of resources | International circulation of capital into and out of New Zealand<br>Standardized treatment of land based production |
| SECTORAL CONDITIONS | Focus on afforestation<br>Attractive for new entrants | Focus on utilisation<br>Pressures for consolidation of forest ownership | Focus on afforestation<br>1962 Forestry Encouragment Loans<br>1965 Current year deductions<br>1971 Forest Encouragment Grant Scheme<br>1980 Cash refunds | Reassessment of conditions suitable for processing and further afforestation |
| STATE INDUSTRIES | SFS develops afforestation | NZFS directly involved in production planning and to a lesser extent processing (eg. Waipa Sawmill) | NZFS expanding State Forests ; co-ordinating Sectoral planning via Forestry Conferences, development of forest processing,marketing and data bases, directing of biological, economic and social research | NZFS disestablished<br>Partial privatisation of State Forests<br>Creation of FC and NZTL |

**Figure 10.2** Accumulation conditions for afforestation.

(Belshaw and Stephens 1932; Jones 1928). A common claim was that a £25 bond invested in an acre of plantation would return £500 on harvesting in 25 years. *Pinus radiata*, a Californian tree, was the species favoured by the SFS and the afforestation companies, mainly because of its rapid growth and availability of seed. But significant areas of Douglas fir (*Pseudotsuga menziesii*) were also planted along with a range of other species.

Work schemes during the Depression of the 1930s meant that the SFS ultimately planted 400,000 acres (161,000 ha) by 1934, concentrated mainly in Kaingaroa State Forest in the central North Island. After 1934, SFS efforts slackened, in the belief that the timber famine would be averted and that forest monocultures were biologically vulnerable. Company afforestation also ceased abruptly in 1934, with the passage of the Companies (Bondholders Incorporation) Act, introduced as a result of dubious business practices to force bond-selling forestry companies to reorganise themselves so that bondholders became shareholders. The important outcome was that additional funds for continued afforestation, much of which had come from Australia, could no longer be raised as freely as before or with so little restraint.

By the late 1950s, New Zealand Forest Service (NZFS)[2] forecasts suggested an annual wood deficit of 30 million cubic feet (850,000 $m^3$) by 1999. In part this was triggered by official projections showing New Zealand's population reaching 5 million by 2000 (in 1992 it is 3.3 million). To meet national requirements, the NZFS proposed an increase of 60 per cent in the area of plantation forest by AD 2000 to give a total exotic estate of 1 million acres (405,000 ha). Although domestic wood needs were the initial trigger, NZFS strategists foresaw expanded timber exports, with the added advantage of reducing the national economic dependence on a narrow range of agricultural and pastoral products (Grainger 1961). In the 1960s, a state target was conceived for forestry of contributing 25 per cent by value of exports by 1999. This second major phase of planting was to be undertaken by both the NZFS and the private sector. The planting was more widely based, with efforts being made to grow high-quality logs as well as pulp wood. These aims were achieved, but the significance of the log export trade in the 1980s suggests all has not gone entirely to plan (Figure 10.2). Analysts predict little expansion in the domestic market, meaning that the increasing volumes of wood coming on stream during the 1990s will have to be sold in the international market.

## Corporatist afforestation, state-owned enterprise and internationalisation

The 1980s opened with a continuation of the debate between agriculture and forestry interests over rural land, albeit at a reduced level of intensity compared to that of the late 1970s (Forestry Council 1980a; 1980b). More importantly, the corporatist alliance struck between the state, major forestry companies, and small growers culminated in the New Zealand Forestry Conference (NZFC) meeting in 1981. This meeting, a sectoral planning exercise following on from earlier efforts in 1969–70 and 1974–75, established a new round of national planting targets, suggested priority planting regions and nominated ideal planting contributions by company, small grower and the state. Future processing options were also considered in detail, since a doubling of wood yields from maturing forests was anticipated from 1990 to 2000.

In many ways the 1981 NZFC meeting represented the elaboration of the forestry formula of the 1930s. The emphasis was on the creation of a resource base. The Working Party on Afforestation (1981) of the NZFC, for instance, asserted that the meeting of afforestation targets was more significant than who actually planted the trees. A new national annual planting target of 43,800 ha was endorsed by the NZFC. This was significantly smaller than the 55,000 ha target set by the 1974–75 Forestry Development Conference. Underpinning the 43,800 ha target was a desire to even out the age–class distribution of the existing forests and rationalise the spatial pattern of forest plantings since 1960.

These two concerns were reflected in the regional planting targets that were arrived at by the NZFC. In comparatively unimportant forestry districts, almost token annual targets (such as 400 ha in Taranaki) were set, while major priority

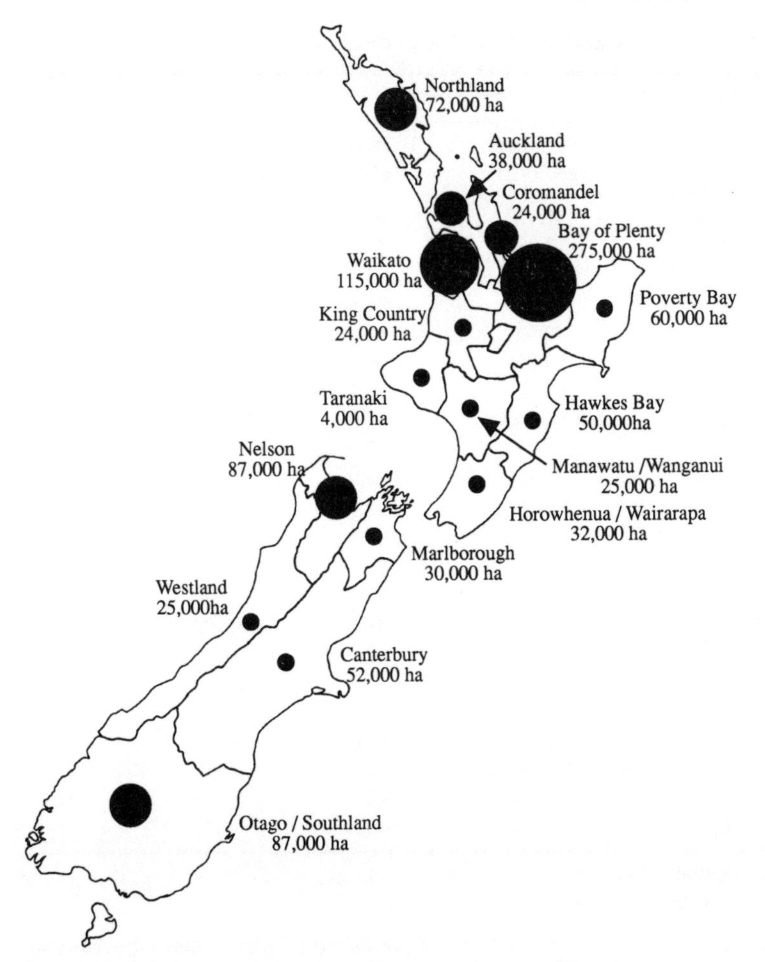


**Figure 10.3** Combined area of state and private plantation forests, 1983.
*Source*: New Zealand Forests Service (1983)

areas such as Rotorua were allocated 5,000 ha for the remainder of the 1980s. The sectoral contribution to regional planting targets also showed wide variation. In Northland, a priority planting region, the state and companies were respectively allocated 42 and 48 per cent of the total target. Nelson, another priority planting area, provides a contrast with the state allocated 57 per cent. In areas such as King Country and Otago, the private sector dominated. Closer attention to planting by smaller timber growers was a feature of the NZFC meeting and signalled that the government was endeavouring to build an alternative non-corporate wood-resource base. Achievement of the regional planting targets rested on sectoral consensus forged at national indicative planning forums. Further real incentives for companies and private growers took the form of tax-law changes to enable a cash refund equivalent to potential reduction in tax liability in 1980 and subsidies in the form of a Forestry Encouragement Grant Scheme which dated from 1971.

**Table 10.1** New Zealand exotic forest estate, 1983

| United Council | State forests | | Private forests | |
|---|---|---|---|---|
| | ha of forest ages 1–25 | 1–25 yr as % of total | ha of forest ages 1–25 | 1–25 yr as % of total |
| Northland | 35,482 | 95.3 | 34,201 | 92.1 |
| Auckland RC | 12,310 | 85.8 | 15,581 | 92.3 |
| Thames Valley | 14,866 | 84.4 | 3,867 | 92.3 |
| Waikato | 10,054 | 89.6 | 67,086 | 82.8 |
| Bay of plenty | 40,532 | 67.8 | 65,286 | 95.6 |
| East Cape | 29,052 | 99.9 | 16,794 | 100.0 |
| Tongariro | 106,334 | 58.2 | 47,826 | 89.8 |
| Hawkes Bay | 24,150 | 68.8 | 16,916 | 96.5 |
| Taranaki | 1,972 | 92.6 | 1,788 | 74.6 |
| Wanganui | 9,505 | 59.0 | 8,128 | 95.5 |
| Manawatu | 396 | 93.8 | 4,800 | 89.8 |
| Horowhenua | 1,961 | 89.1 | 1,870 | 96.2 |
| Wairarapa | 8,361 | 90.7 | 8,757 | 93.6 |
| Wellington RC | 276 | 68.3 | 3,859 | 84.4 |
| Nelson Bays | 39,305 | 93.2 | 28,890 | 94.0 |
| Marlborough | 9,106 | 97.9 | 14,643 | 94.2 |
| Canterbury | 22,219 | 87.7 | 12,915 | 85.3 |
| Aorangi | 4,306 | 99.8 | 5,224 | 67.4 |
| West Coast | 17,448 | 94.6 | 4,084 | 99.8 |
| Coast/North Otago | 7,287 | 97.2 | 6,547 | 92.4 |
| Clutha/Central Otago | 26,589 | 88.2 | 9,642 | 88.4 |
| Southland | 11,004 | 90.6 | 9,254 | 89.7 |

RC Regional Council
*Source*: Butler *et al.* (1985)

Both these incentive measures were abolished by the Labour Government in 1984 in favour of a 'neutral' tax regime which sought to facilitate investment decisions on the basis of real profitability rather than by tax advantages.

By 1983 the second planting boom was peaking: sizeable new forests had been created throughout the country (Figure 10.3). The Central North Island remained dominant, but important secondary areas existed in Northland, Nelson, and Otago/Southland. Almost all the state forests outside central North Island were established during this second planting boom (Table 10.1). A similar pattern holds for private forests, but is not so clearly defined, reflecting attempts in earlier years to undertake company afforestation and create farm woodlots throughout the country.

As the 1980s unfolded, the NZFS, the major afforestation organisation, came under increasing scrutiny from a number of directions, some of them unexpected. An antagonistic relationship between environmental groups and the NZFS had emerged and gathered momentum since the mid-1970s. The initial point of contention was NZFS plans to use two sizeable areas of indigenous beech (*Nothofagus* spp.) in the South Island to feed a proposed pulp-mill operation.

Ultimately the 'Beech Scheme' lapsed, but the concerns of environmentalists persisted over what they saw as the NZFS's inability to understand and accept forest preservation as a desirable goal for all remaining indigenous forests. The replanting by NZFS of cut-over forest with exotic species, although not undertaken on a large scale, further aggravated this conflict. The afforestation policy of NZFS more generally during the 1970s and early 1980s also disturbed sections of the rural community (Smith 1981). Higher planting targets prompted rural concern that lands then devoted to agriculture might ultimately be converted to forests either by NZFS or corporates purchasing land at a time when the rural land market was vulnerable. Throughout the 1970s, the economic rationales for plantation location and size also became eroded as forestry became incorporated into regional-development strategies under the Labour Government (1972–75), and as tree planting was increasingly used to mop up seasonal unemployment and/or to provide jobs in economically depressed districts. The reservations of forestry professionals that these strategies would lead to forests that might never be profitably harvested were brushed aside (Grant 1979).

From the late 1970s there was growing criticism of the role of the NZFS as a forest grower, a wood seller, and a sawmill operator. The internal reconciliation of goals of production and protection forestry within a single department was attacked by environmental groups. From a production perspective there were also signs of a reappraisal by 1978. The critical event was the investigation of the Parliamentary Public Expenditure Committee into financial management in the NZFS. The sub-committee's report (the McLean Committee) suggested that although the NZFS cash-accounting system gave parliamentary accountability, it was not organised in terms of profit or loss. In essence it was a managerial and not a commercial accounting system. Given the growing role of forestry in the economy, the lack of flexibility and of operational independence was highlighted by the sub-committee as especially significant. The completed report was not tabled in Parliament until 1980 and there was little immediate political response (*AJHR* 1980, I12A). Separately, in 1980 the Development Finance Corporation's Forest Industry Study commissioned by the NZFS reinforced the McLean Committee's findings. Specifically, the Forest Industry Study highlighted a lack of clarity in government objectives and priorities for the forestry sector and the role of the NZFS within the sector (Development Finance Corporation 1980).

Over the next few years a variety of organisational forms were considered. The NZFC in 1981 provided a forum for discussion, but did not feed into a ministerial proposal in 1982 to merge NZFS and the Department of Lands and Survey. In 1982 the government reviewed the objective, functions, and activities of the NZFS, but largely supported the status quo (State Services Commission 1982). This included favouring a single organisation for managing indigenous and exotic forests and a rejection of the sale of exotic state forests and separation of state sawmilis from NZFS. However, separate commercial and forest servicing divisions and a Board of Management were proposed.

A change of government midway through 1984 accelerated the pace of change, largely through an alliance of environmentalists (seeking removal of control of indigenous forests from NZFS control) and the Treasury which challenged the efficiency of NZFS as administrators and as a wood grower. The outcome in 1987, with a number of possibilities being explored along the way, was the disestablishment of the NZFS, with forest preservation being vested in a newly created Department of Conservation, while production forests (almost exclusively exotic plantations) were transferred to a New Zealand Forestry Corporation (FC). A Ministry of Forestry was set up to fulfil policy and regulatory requirements.

The demise of the NZFS marked the end of a corporatist era of collective sectoral planting targets and planting programmes. The FC was established as a State-Owned Enterprise (SOE) – a publicly owned business which was charged with selling wood on commercial lines, with no regard to the sectoral imperatives that had influenced earlier NZFS planting regimes and wood sales. Wider government moves in the period from 1984 to 1987 further signalled an unprecedented reworking of the state–economy interface, both domestically and externally. Subsidies, incentives, and tax concessions covering land-based production in New Zealand were stripped away in a state-led effort to replace economic regulation by market liberalisation. The impact was immediate: annual planting rates dropped steadily from 54,000 ha in 1984 to 13,000 ha in 1991.

A commercial mandate for the FC also signalled that some existing timber sales agreements would be renegotiated. The Tasman wood sale of the mid-1950s which supplied a joint state–private sector pulp and paper mill at Kawerau, for instance, was set up under conditions that were very attractive to the wood buyer, Tasman Pulp and Paper Company (TPP) (by the 1980s it was no longer a state-linked concern). However, before FC was far along the pathway of a separate existence as an SOE, it suffered three blows. The first of these occurred when the courts ruled that SOEs would not be able to sell former Crown lands which they now owned, but which were subject to claims before the Waitangi Tribunal.[3] Secondly, forest valuation is a complex issue at any time (e.g. Fraser *et al.* 1985), but FC and Treasury estimates of the value of the plantations that the corporation was to purchase from the government were at variance, even allowing for strategic negotiation: $1 billion (FC) versus $4 billion, later $7 billion (Treasury). Before settlement was reached the Labour government began, in 1988, an extended exercise in privatising a range of state assets. In July 1988 the government announced the sale of 550,300 ha of state plantation forest and the two state-owned sawmills. The sale was to be handled by FC, which was transformed from a forest management organisation to an asset sales unit. Labelled by some forestry commentators as the 'sale of the century', this attempt to sell over 500,000 ha of forest represented an unprecedented chapter for plantation forestry in New Zealand.

## Restructuring state forestry

As part of the sale process, the government appointed a special working party to devise guidelines which would maximise revenues and facilitate the sales without breaching government policy with regard to the principles of the Treaty of Waitangi or environmental conservation practices. The revised sale procedure involved selling the trees but not the land by granting a Crown Forest Licence for 70 years (or two forest rotations) to successful tenderers. The fuller rationale for the forest sale as detailed by FC's Chief Executive was fivefold. The basic proposition was that the forests of the second planting boom had always been designated for export. Since these forests were nearing maturity it was agreed that a transfer of cutting rights to the private sector was timely. Such a transfer was desirable because expanding the forest-processing sector to meet increased wood harvest was estimated at costing $7 billion and the state had no desire to move into this area. Two further linked justifications for the sale were that outright sales would return many years of public investment in forestry to the taxpayer, and that it would meet the government's debt-reduction programme (Kirkland 1990). The actual sales were decided by tender. However, the New Zealand Commerce Commission declined blanket coverage of bids for all forests by the main forestry companies FCL, NZFP and CHH on the grounds that they already had a dominant position in some regions. Local registrants predominated but Australian, Chinese, Hong Kong, Japanese, US and Korean firms were all represented. Interest from overseas forest-owning and forest-processing firms was limited principally to ITT Rayonier of the USA and Chonju Paper Manufacturer of Korea. Most of the overseas interest came from what might broadly be described as trading and investment companies, some with only limited previous involvement in forestry.

Eight firms purchased, by tender, 22 forests (135,622 ha) and one sawmill for $588 million. Over half of the firms were based in New Zealand, and they secured 45 per cent of the forests and 40 per cent of the area sold. These firms included major New Zealand companies such as FCL subsidiary Tasman Forestry (48,852 ha) and CHH (7,504 ha). Three small local companies acquired minor amounts. The new entrants from overseas were Juken-Nissho Ltd (43,531 ha), a Japanese joint-venture company, Wenita Forestry Ltd (20,521 ha), a subsidiary of Chinese–Hong Kong firms, and Ernslaw One Ltd (23,801 ha), a subsidiary of Singaporean and Hong Kong interests (Roche 1991).

All tenders for the prime central North Island forests were rejected. Thereafter a round of negotiated sales began which resulted in CHH paying $313 million for 21 forest blocks totalling 94,121 ha in Canterbury and Hawkes Bay. This purchase enabled CHH to secure wood supplies for its joint-venture Whirinaki pulp mill in Hawkes Bay and its Canterbury board mills. At the end of 1990, the National government replaced Labour and halted the sales: 43 of the 90 forests had been sold, amounting to 239,881 ha or 44 per cent of the extent. Since then, further changes have taken place. ITT Rayonier, an unsuccessful bidder for state forests, has purchased the 650 ha Kohitere plantation near Levin from the Department of Social Welfare. More significantly, CHH, as part of a complex set of financial transactions, which saw it purchase Elders Resources New Zealand Forest Products' (ERNZFP) 141,021 ha plantation forest and processing complex in 1991, has reduced its share in the Whirinaki pulp mill to 10 per cent and turned over its Hawkes Bay forests to the joint-venture company Carter Oji Pan Pacific Ltd (COPP).

The 'market mechanism' has not delivered what the politicians promised as far as the sale of state forests is concerned. The average price of the forests sold by tender was $4,163 per ha, translating to about $2.3 billion for the entire 500,000 ha, compared with the Treasury's $4 billion and later $7 billion estimates. Also, not all the forests were sold, so that the attempt to quit direct state involvement in production forestry can be regarded as a failure. Over 163,500 ha remain under the control of reconstituted SOEs such as N.Z. Timberlands (Bay of Plenty) Ltd (NZTL) and N.Z. Timberlands (Westland) Ltd.

Changing ownership is only part of the picture. Given that fears were voiced before the forest sales programme that it would provide the opportunity for overseas interests to engage in clear felling and log exports to the detriment of domestic environmental and economic concerns, investment by new entrants in processing and replanting needs to be scrutinised. Whether new forest owners would actually replant was the subject of some debate. The philosophy of the Forestry Working Group (1988) that in order to maximise revenues, cutting rights should as closely as possible approximate rights of freehold caused some concern. The FC argued that the level rental charges for the Crown Forest Licences which would facilitate the privatisation of the forests would also act as a strong in-built incentive to encourage replanting. The incoming National government late in 1990 favoured mandatory replanting. This stance on replanting was reiterated in July 1991, at the time of the decision to resume the sale of unsold state forests (excepting Bay of Plenty and Westland). The only exception was to be that a particular area could be converted to some other sustainable use. Official forest-sales guidelines made public late in 1991 place compulsory replanting at the top of the list (*Evening Standard*, 25 October 1991).

## Local integration of New Zealand's plantation forests into global wood networks

During the 1980s, leading New Zealand-based forestry companies such as FCL, CHH and, to a lesser extent, NZFP all internationalised their operations, purchasing forest and processing plants or setting up joint ventures in countries such as Australia, Canada, USA, Chile, and Brazil (Le Heron 1988). Export of New Zealand-produced forest products was always part of these firms' activities, so that the transition from nationally based export traders to internationalising companies operating in a globalising market was not an unexpected trajectory, especially as they were outgrowing the New Zealand economy. The opening of the New Zealand economy to international market forces by the Labour government from 1984 to 1990 was the precursor to further internationalisation in forestry, involving the incorporation of local plantation-forest resources into global wood networks. To an extent this was already taking place before the state forests were put up for sale. For example, ITT Rayonier had already established an office in Auckland and was purchasing small private forests across New Zealand in order to engage in the export log trade. More conspicuously, NZFP, in a complicated series of manoeuvres, ended up being taken over by Australian agri-business firm Elders IXL, through its investment arm Elders Resources, until its acquisition by CHH in 1991 as a consequence of financial troubles and restructuring of Elders. Without doubt the forest-sales strategy has greatly hastened the integration of local wood supplies into global networks by introducing new links with Singaporean, Hong Kong and Japanese interests. However, the actual fashion in which the 'local' is connected with the 'global' varies in the different forest-growing regions of New Zealand. Something of this variation is sketched in the next section.

## Forest-growing regions

### *The core region*

The major plantation forestry region of New Zealand is situated in central North Island. Initial planting began in 1898 and the main SFS and afforestation company efforts were concentrated here in the 1920s. The flat or rolling country, covered by grass and scrub, was easy to plant and large amounts were Crown land. Importantly this land was not sought by agricultural interests as it suffered from a then unrecognised trace-element deficiency which caused stock to waste away and die from 'bush sickness'. In the 1950s, large-scale utilisation began with the establishment of a state-sponsored pulp and paper mill at Kawerau drawing on Kaingaroa State Forest (114,928 ha), then the largest plantation in the world, and the NZFP pulp mill at Kinleith near Tokoroa, central to that company's forest estate. Both NZFS and NZFP continued to expand their central North Island forests during the second planting boom and both NZFP and TPP extended their pulping operations. In 1971 the last sizeable Kaingaroa tender (54,280 $m^3$) was awarded to a new consortium, COPP, which constructed a further pulp mill some distance away at Whirinaki near Napier in Hawkes Bay (Figure 10.4). The New Zealand firm Carter Holt Ltd (a recent amalgamation of two old-established regional sawmilling companies) and Oji Paper of Japan were behind this venture.

A further international presence was felt in 1984 when NZFP came under Australian control. Previously FCL and CHH had both expanded offshore. Subsequently CHH has taken control of NZFP but restructured its joint venture with Oji Paper Ltd to a 10 per cent holding as well as transferring all the Hawkes Bay plantations it purchased from FC to COPP. In addition, corporate raider Brierley Investments Ltd, which had built up 32 per cent of CHH, has sold half of this stake to the US firm International Paper Ltd for $454 million, as part of a joint-venture shareholding in CHH. International Paper Ltd, with an annual turnover of $22 billion and dealings in 120 companies, sees the CHH as its 'window into greater involvement in the Pacific Rim' (Hargreaves 1991).

Kaingaroa, the largest forest, remains in state management under NZTL. The future impact of global pressures on the central North Island forests and their owners may be significant in the short to medium term. For example, the existing pulp and paper facilities were originally developed in the 1950s: TPP's performance will be compared with FCL's other pulp and paper operations in the USA, Canada and Latin America when decisions about reinvestment in forest processing are made. This is not to predict a company pull-out, but rather to emphasise that investment decisions will not be limited to those of national economies and export markets. Should there be a share float for NZTL, there would also be multiple implications for the future of existing log sales (TPP, for instance, was established on the basis of three 25 year sales – initially with wood very attractively priced from the company perspective). Will the mix of shareholders jeopardise or continue these contractual arrangements? Will NZTL remain a wood seller or engage in more processing, or will it continue to sell on an attractive log-export market because that brings favourable finance returns to the new company, at the expense of value-added forest processing? Furthermore, will

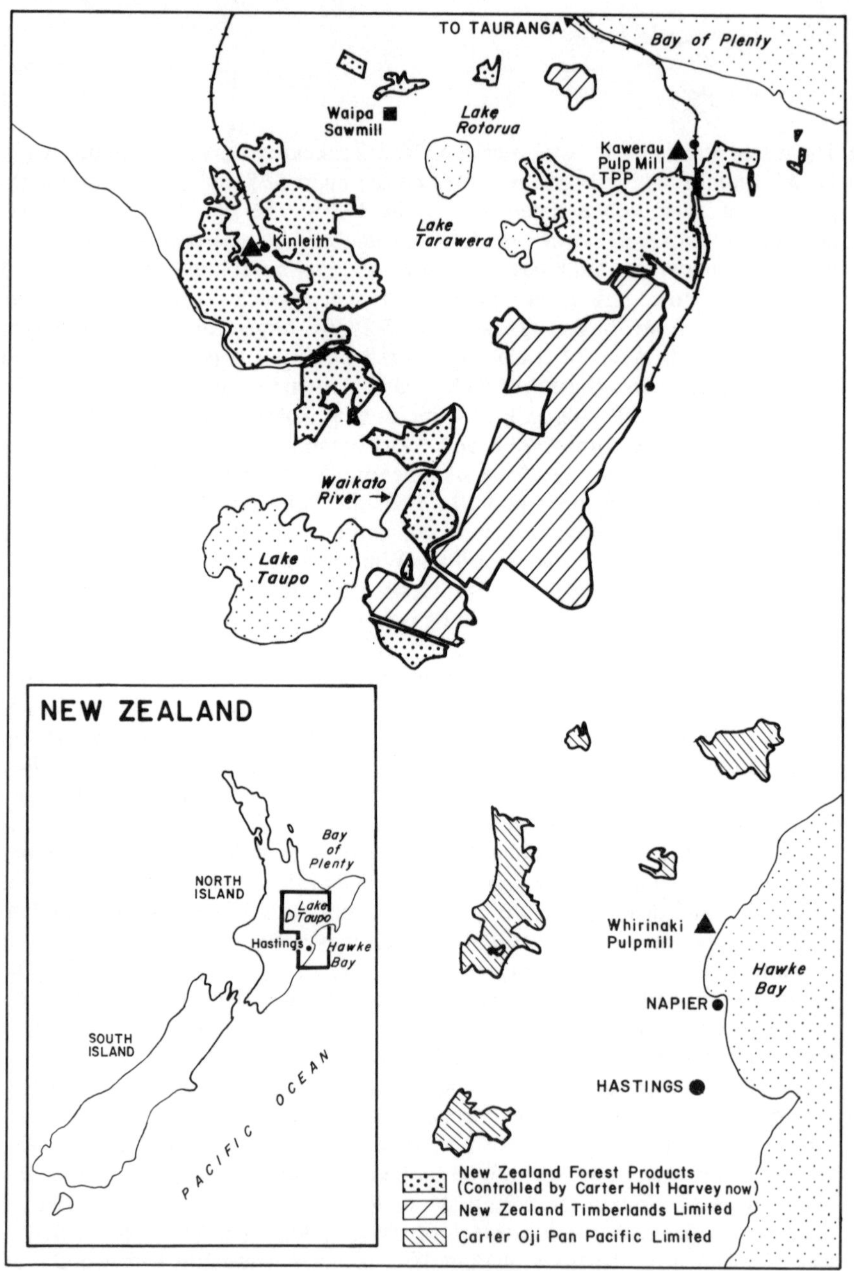


**Figure 10.4** Recent ownership changes in the core forestry region.

shareholders in NZTL see it as a long-term operation or will they go for one-off realisation of profits from wood sales? Perhaps NZTL may decide to participate in a 'third planting boom'. Add the likelihood of sizeable overseas shareholding to the picture and the range of possible outcomes is great. For New Zealanders employed in the forestry sector in this region, the introduction of new milling technology, coupled with tree-growth rates and improved silviculture will probably see the profitability of the industry maintained, but it may not mean an expansion in employment opportunities. Likewise, more direct links to international markets may also mean the impacts of fluctuations in prices of forest products will be translated more quickly and completely to the region, so that afforestation and production levels will be adjusted more abruptly up and down than was previously possible or desired.

### *Otago*

Otago was the site of considerable afforestation experiments and actual planting efforts before the 1920s, in part to meet anticipated timber demands in a comparatively treeless region and because there was Crown land available. The region is also notable from a forestry viewpoint for the early plantings of the Dunedin City Council. In the 1950s the NZFS established its second major sawmill at Conical Hills near Tapanui State Forest, where it became a major employer in the district. Otago was given 'priority planting' status during the second planting boom, and both the NZFS and forestry companies acquired sizeable land banks (7,232 ha and 2,355 ha, respectively, by 1981) in the district (Working Party on Afforestation 1981). The acquisition programme, as in other regions, raised concern in the local farming community about corporate land owners and forestry as a competing land use (Aldwell and Whyte 1985).

Once the programme of forest sales was announced a number of local and overseas companies tendered for forests in Otago. These included ITT Rayonier, which was already buying logs in the area, and the successful new entrants Wenita Forestry Ltd and Ernslaw One Ltd. A joint venture between Togen Ltd (Hong Kong) and Sinotrans (Chinese) Wenita Forestry Ltd purchased Berwick (11,282 ha) and Otago Coast Forests (9,239 ha) for $115 million. Ernslaw One Ltd secured the 10,615 ha Tapanui forest and the Conical Hill sawmill. The latter is controlled by a subsidiary called the Blue Mountain Lumber Coy.

Wenita Forestry Ltd quickly entered the export log trade, sending its first shipment of 24, 000 tonnes of *Pinus radiata* logs to Korea in 1991 (*Otago Daily Times,* 23 February 1991). Ninety per cent of the Wenita's logs are exported, a move necessary in order to recoup returns and one which has led to an increase in forest harvesting. The company's longer-term plans include local processing, probably in the form of a plywood mill for export and replanting of harvested areas (*Otago Daily Times*, 8 February 1991). Wenita's involvement in Otago clearly reveals some aspects of the globalisation of regional forest resources in New Zealand. Their log sales and projected plywood sales are mainly overseas, principally in Asia. Neither Togen nor Sinotrans is predominantly involved in forestry, and Togen has hinted that future investments in New Zealand may be in other sectors such as food and coal mining, and presumably in other parts of the country (*Evening Post,* 7 October 1991).

Ernslaw One Ltd successfully tendered for six forests in both the North and South Islands. A diversification strategy may underpin this pattern. Like Wenita

Forest, Ernslaw One quickly entered the log export trade dispatching Ponderosa pine (*Pinus ponderosa*) and *Pinus radiata* to Korea, with a view to extending to other markets. Most of the logs came from areas of the forest planted in the 1930s and now past their prime. An anticipated 144,000 $m^3$ of logs will be exported in 1991 to 1992 (*Otago Daily Times*, 10 May 1991). The company intends replanting felled areas and expanding sawn-timber sales to Australia, a traditional market for Conical Hill timber. For local West Otago communities, the mill sale has been significant. It is the largest employer in the district and initially fears were held that new owners might decide to close the operation down.

Ernslaw One integrates Otago wood supplies into global markets in some ways which are similar to Wenita Forestry, particularly with respect to log exports. However, Ernslaw One brings another link, this time with tropical forestry via its parent companies, Habacus Pte Ltd of Singapore and Shiang Yang International of Hong Kong, which have logging and processing operations in Papua New Guinea. Another subsidiary, Rimbunan Hijau Sdn Bhd, owns a large Malaysian plywood plant and holds a logging concession in Sarawak. Limited tropical hardwood supplies, plus threats to the long-term future of tropical logging, have persuaded Ernslaw One Ltd to invest in New Zealand softwood plantations.

## Towards a third planting boom?

The Labour government, having implemented a major reorganisation of environmental administration in 1987 and initiated the sale of state forest assets in 1988, then readdressed the question of defining a forest policy different from the balanced-use policies of the 1970s and 1980s. The MoF itself in 1989 sought to redefine forest policy which would 'identify primary or predominant uses as the way to go into the 1990s' (Valentine 1989) with a more market-oriented forestry environment. A Ministerial advisory committee of public and private-sector foresters as well as environmentalists moved in this direction by issuing a forest policy discussion document in August 1990. The document unequivocally stressed the need for a national forest policy, on the grounds that a clear policy could be more effectively implemented, that it provided for long-term continuity, and that it facilitated effective management decisions by officials. The forest policy discussion document was organised around 'general principles' and 'desired outcomes' – the new structure for public service operations introduced by Labour in 1985. Two general principles advanced by the advisory group are of interest in that they do challenge the 'more market' philosophy: firstly, the need to account for non-market values and market imperfections; and secondly, that market and non-market benefits need to exceed their costs, and the costs should be borne by those who benefit. Translated to desirable forestry outcomes as far as they are related to afforestation, sustaining site quality, and local or central government assistance for finance, planting and management of forests were warranted 'where the market fails to meet socially desired goals' (Advisory Group 1990). The rationale for renewed government intervention also included situations where there was a continuing need to reduce soil losses from farmland, where there was a need to increase carbon storage in trees to mitigate the build-up of atmospheric carbon dioxide, to provide 'insurance' plantations or domestic wood requirements to guard against catastrophic losses of existing *Pinus radiata* plantations, and finally to include a diverse range of special-purpose trees amongst the three preceding areas.

Two other desired outcomes related to the design, planting and management of forests as well as increased employment and national and regional development. Specific actions required to facilitate the above included a revised taxation regime where expenses are carried forward, and an easing of some of the restrictions imposed on forestry as a land use under the Town and Country Planning Act 1977 (Meister and Fowler 1984).

A change of government, back to National, late in 1990 brought the emerging forest-policy strategy to a halt. The difficult issue to resolve is whether by 1990 the MoF was looking to a forest policy because of perceived shortcomings in 'more market'-based forestry strategies, or because a forest policy *per se* provides a justification for agencies such as itself. At this stage it is difficult to offer a definitive answer. Suffice it to say that by May 1991 a new working group was assembled to investigate joint-venture forestry agreements as part of a concerted push by National's Minister of Forestry. The Minister's message was hardly new (his strategies have their antecedents in the first and second planting booms), but whether direct and indirect incentives will have the impact of those of the 1920s and 1930s, and of the 1960s to 1980s, remains at best problematic and arguably rather unlikely. A central aspect of concern was the decline in annual afforestation rates from 54,000 ha in 1984 to 13,000 ha in 1991. Log exports have risen, approximately fourfold, during this same era (Figure 10.1). The National government's emerging forest policy stresses a return to earlier expanded annual planting levels to encourage more domestic processing and hence the value-added component of wood exports.

The setting of a new planting target for the period from 1991 to 2020 is to use a proven device from earlier planting booms. The newly announced target is every bit as bold as that of 1925, calling for 100,000 ha per annum (or double the maximum rates of the second planting boom) and continuation of this effort for a full 30 years (i.e. a complete rotation). If attained, this target equates with 4.2 million ha of plantation forest by 2020, or 15.6 per cent of New Zealand's land area. Again, following the model of the 1970s and 1980s (Thomson 1975; Kirkland 1981), the MoF (1991b) suggested that this land is available in the form of marginal farmland and would be better utilised under forests. The statement that 'there is no economic reason to limit a forestry investment to the marginal areas of a farm or to low quality land' (MoF 1991b) again echoes an emergent forestry-sector position of the mid-1970s (Le Heron and Roche 1985). A new era of land-use debates in the 1990s will proceed along different lines from that of the 1970s, not only because of the partial exit of the state from sectoral intervention and forest ownership and the entry of internationalised forestry capital, but also because a new omnibus Resource Management Act 1991 is now in force. The Resource Management Act has replaced the Town and Country Planning Act of 1976 and more than 50 other statutes. Under the new legislation the effects of a land use are what is critical. The privileged position that agriculture enjoyed under the 1976 Act, where the protection of land of high actual or potential capability for the production of food was paramount, has been removed. Since the environmental effects of forestry potentially may be less than those of agriculture, the new planning legislation presents a more 'level playing field' than forestry faced in the 1970s.

The 100,000 ha planting target differs, however, from those of the 1960s to 1980s in several important respects. Firstly, it has no identifiable state-planting programme at its core. Secondly, no detailed regional planting targets have been expounded. This is, at least, in keeping with the 'more market'-oriented philosophy which argues that forestry investors relying on market signals will create optimal forestry configurations superior to those worked out in the sectoral planning exercises of the 1960s and 1970s. Thirdly, it is not envisaged that forestry companies alone will meet the target. This shortfall is to be met by encouraging joint ventures between farmer land owners and private investors to create further plantations. A number of legislative and financial instruments for encouraging this sort of afforestation activity already exist, having been developed in the early 1980s when the NZFS was endeavouring to stimulate the creation of a small-grower wood resource in order to offset the dominance of forest ownership by large companies. The MoF (1991b) has recommended a range of taxation charges, along with the repeal of Labour's 'cost of bush' formula for new planting which delayed deductibility of forest-crop costs until the sale or harvest of the forest, in favour of a return to the earlier annual deductibility of direct planting costs. The ratio of company to private-investor planting rates is unspecified, but a considerable number of small-scale private forestry schemes have been soliciting investors since 1991. An example of the latest small-scale private forestry ventures is the Arapaoanui Forest Partnership which proposes to plant a 285 ha block 32 km from Napier in Hawkes Bay. The compound rate of return after tax by AD 2021 is given as 8.33 per cent (Arapaoanui Forest Partnership 1992). Overall, the MoF has estimated an aggregate new planting rate in 1991–92 of 31,000 ha.

The present Minister of Forestry, unlike his predecessors, does not have a department which is organised in such a way as to plan and undertake a large-scale national afforestation programme. With no direct involvement in tree planting by government agencies, is the Minister of Forestry's target of 100,000 ha per annum realistic? Put another way, will private-sector investors plant, and at the scale that is being advocated? Taxation changes and other alterations to the investment environment may in effect represent a tilting of the 'level playing field' sufficient to encourage accelerated afforestation rates. In the absence of earlier incentives such as the Forestry Encouragement Grants and direct NZFS planting, it requires considerable optimism to see the current initiatives producing a third planting boom of more than twice the magnitude of earlier afforestation eras.

There are signs that the new overseas entrants are investing in local processing. Juken-Nissho, for instance, is constructing plywood mills at Carterton and Gisborne close to its forests. The $40 million Masterton mill is expected to process 100,000 m$^3$ of log annually, producing plywood and employing 120 people in the mill and another 60 in forestry and transport-related activities (*Evening Post,* 10 April 1991). Planning permission has been sought for a $40 million Gisborne mill and a further mill is scheduled for Northland in the mid-1990s when Aupouri forest is ready for harvesting. Furthermore, Juken Nissho and Ernslaw One Ltd are already replanting their forests, allaying initial concerns that the overseas purchasers would look to a one-off harvest of their forest purchases and and to subsequent log exports. However, it seems doubtful whether the new entrants from overseas will play an especially large part in any third planting boom. Though they have made purchases, the next priority will be to realise a return on the investment, which presently involves participation in the log export trade and will include processing once the new mills are built. Although 180 jobs in Carterton (population 3,902 in 1986), for example, is significant locally, it needs to be traded off against the closure of the local sawmills that previously handled wood from state forests in the area. Furthermore, the new processing plants will doubtless employ sophisticated technology as demonstrated by two other recent mills in which Japanese companies have an interest. Tachikawa Forest Products Ltd has built a new $20 million computerised precision-cutting sawmill at Rotorua which will produce annually 36,000 m$^3$ of 12–14 mm by 120–240 mm thin board for packaging and employ 70 staff (Edwards 1989; *Sunday Star*, 9 July 1989). The Nelson Pine Industries Medium Density Fibreboard mill near Nelson, a joint venture between Corporate Investments and Sumitomo Forestry Corporation of Japan, has a 200,000 m$^3$ capacity, making it one of the largest single-site plants of its type. The doubling of capacity in 1991, five years after the mill was built, cost $60 million and will only give 35 new jobs (*Dominion*, 31 October 1989). The consequence of this is that a forestry-products boom, with New Zealand-grown wood being more widely sold on international markets, would probably not be associated with any comparable growth in forestry-related employment in New Zealand.

The purchase and afforestation of additional areas is, of course, a possibility, but whether the new entrant firms will do so, either directly or via joint ventures, is difficult to ascertain. In part, this is because of the internationalised context in which decisions about further afforestation and processing activity are to be made, involving the competitiveness of New Zealand as an arena for forestry investment. At the New Zealand end, any discussion of expansive tree planting by overseas-based forestry companies which involves further land purchase will be debated vocally in the public sphere (the issue of foreign ownership of land is one of long-standing sensitivity), and in an institutional setting in terms of local-planning regulations and consents. Finally, the very notion of a forest-planting target springs from a particular view of the nation state – one from which the reality of an internationalised forestry sector in New Zealand is far different. Will overseas companies feel any need to respond to the New Zealand Minister of Forestry's planting target? More importantly, will New Zealand-based firms such as FCL and CHH react any differently?

The basic consideration is the extent to which FCL and CHH need or desire to extend their forest holdings further. Their forest resources are already extensive: prior to the state forest sales in 1989, FCL held approximately 121,935 ha or 11.5 per cent of the total plantation forest area. CHH at this stage had approximately 55,136 ha or 5.2 per cent, with NZFP owning another 141,021 ha or 13.3 per cent of the forests. From the sales programme, CHH directly or indirectly added another 101,634 ha and FCL 48,852 ha. With its subsequent acquisition of NZFP, the total group forest area is now close to 300,000 ha. Under these circumstances, it is possible that the major forestry companies will have an interest in expanded afforestation (as opposed to replanting of harvested plantations) only where it improves the configuration of their existing forests. Besides, as internationalised firms, they will also have to take account of numerous considerations, some external to New Zealand. To complicate the picture further, the core central North Island forests were unsold and were subsequently placed under the control of NZTL. Since NZTL may itself be subject to a share float, in the manner of the Bank of New Zealand, Air New Zealand and Telecom, it is possible that future corporate wood demands may be sourced from a NZTL shareholding rather than an expanded afforestation programme.[4] With wood harvesting increasing from 1990 to 2030, processing may for the time being also rank higher in corporate considerations than supporting a new national planting target.

The most likely supporters of the new planting target in the early 1990s are small private investors and farmer joint ventures encouraged by changed taxation regulations. The possibility of additional joint ventures involving new-entrant Japanese forest does exist, given the willingness of Japanese companies to operate in this fashion. However, given that the area planted under the Forestry Encouragement Grant Scheme, which was much more attractive than the existing incentives, only totalled 100,000 ha for the period from 1970 to 1983 (*AJHR* 1991, C16), the target seems unlikely to be met. The spatial patterning of planting is left unconsidered by the policy makers. There seems to be little guarantee that what is planted will be optimally located with regard to export markets.

## Conclusion

The restructuring of the state sector in New Zealand during the mid-1980s led to the dissolution of the NZFS, a major contributor to early afforestation efforts in New Zealand. The much narrower policy-regulatory role of the MoF and the commercial mandate of the FC means that there is no state afforestation agency in existence nor can one be easily created under prevailing approaches to the role of the state sector.

In the mid-1980s the long-standing corporatist alliance between state and forestry companies to encourage, target and assist afforestation efforts also came to an end. Although MoF is trying to 'talk up' afforestation rates it is doubtful whether comparatively modest taxation adjustments will provide sufficient incentive for a take-off in planting rates. The targeting of joint ventures between private investors and farmer land owners will probably see a significant number of new small forests, but of itself seems hardly likely to contribute significantly to a 100,000 ha annual target. Finally, the commercial decision-making environment is much more complex – even in the regions. Local New Zealand forestry plantations are much more strongly tied into global wood systems than before the sale of state forests. This has two major implications for afforestation. Firstly, the decision by an international firm on whether it will reafforest its New Zealand land, or afforest new land acquisitions, is situated among a raft of more immediate processing investment requirements and other competing investment alternatives in other parts of New Zealand and offshore, and both within and beyond forestry. Secondly, additional rounds of forestry investment will probably emphasise technology, so that jobs in the regions need not flow from the establishment of profitable forestry operations.

This chapter portrays the capitalist nature of afforestation in New Zealand, an experience which can be situated in the wider developments of international forestry. New Zealand has become progressively more integrated into the emerging global economy, firstly by way of exports, and secondly through financial flows related to ownership of forests and processing plants. This integration, explored at the national and local level, is probably underway in other countries as the core industrial economies, such as Japan, seek to develop alternative sources of raw materials. A research agenda for the 1990s is the exploration of the global forest economy from a perspective which centres on identifying the incorporation of all aspects of industrial forestry in the world capitalist-accumulation process. This agenda would provide a basis for explaining investment decisions in national contexts which have been made by economic agents whose sphere of operations is often multi-national.

## Notes

1. The pattern is difficult to reproduce meaningfully at a large scale. Detailed maps of the 1840 (pre-European) and contemporary forest cover of New Zealand are to be found in A.H. McLintock's *Descriptive Atlas of New Zealand* (1959) and I.M. Ward's *New Zealand Atlas* (1976). In addition, the NZFS FS Map Series 2 at 1:1,000,000 scale indicates state and private forests.
2. The State Forest Service was renamed the New Zealand Forest Service under the Forests Act 1949.
3. The Waitangi Tribunal was created by the Waitangi Tribunal Act 1975 to examine the main grievances, particularly those relating to land purchase. The Labour government's (1984–90) Amendment Act in 1985 allowed the Tribunal to consider cases stretching back to the signing of the Treaty of Waitangi in 1840. Under the terms of this Treaty, Maori chiefs ceded sovereignty to Britain but retained 'Full exclusive and undisturbed possession of their Lands and Estates, forests and fisheries ... so long as it is their wish and desire to retain the same'. The Waitangi Tribunal has been called on to deliberate on breaches of the principles of the Treaty of Waitangi. Although the Tribunal is empowered only to make recommendations to government the weight of evidence in its reports meant that the Labour government paid close attention to Tribunal recommendations.
4. On 30 April 1992, after this chapter had been written, the sale by tender of a further 34 forests totalling 97,453 ha was announced. The successful bid of $NZ 366 million was made by the US company ITT Rayonier, and netted forests spread across both North and South Islands. This development strengthens the US presence to the extent that there are now discernible Asian and North American investment axes in forestry in New Zealand. The major Bay of Plenty forests remain unsold, still beset by legal disputes over long-term wood-supply contracts to Tasman Pulp and Paper Co. and more recently with claims for Kaingaroa and Rotoehu forests being made to the Waitangi Tribunal.

## References

Advisory Group Reporting to the Minister of Forestry (1990) *A forest policy for New Zealand*. Ministry of Forestry, Wellington.

AJHR (*Appendices to the Journals of the House of Representatives*) (1925) C3; (1980) I12A; (1991) C16.

Aldwell, P.H.B. and Whyte, J. (1985) Impacts of Otago forest sector growth in Bruce County, Otago: a case study, *New Zealand Journal of Forestry* 29(2): 269–95.

Arapaoanui Forest Partnership (1992) *Forestry investment*.

Belshaw, H. and Stephens, F. (1932) The financing of afforestation, flax, tobacco and tung oil companies, *Economic Record* 8: 237–59.

Butler, C., Levack, D., McLean, D. and Sharp, J. (1985) *A national exotic forest description system*. Forestry Council, Wellington, Working Paper No. 3.

Development Finance Corporation (1980) *Forest industry study*. Development Finance Corporation, Wellington.

Edwards, V. (1989) Tachikawa joint venture will supply Japanese market, *N.Z. Forest Industries* 20(10): 44.

Forestry Council (1980a) *Forestry as a land use*. Bulletin No. 6, Forestry Council, Wellington.

Forestry Council (1980b) T*he place of forestry in the Wairoa district scheme*. Bulletin No. 5, Forestry Council, Wellington.

Forestry Working Group (1988) *Sale of the Crown's commercial forestry assets*.

Fraser, T, Horgan, G.P. and Watt, G.R. (1985) Valuing forests and forest land in New Zealand: practice and principles, *FRI Bulletin* no. 99, Food Research Institute, Rotorua.

Grainger, M.B. (1961) The future demand for forest products in New Zealand, *New Zealand Timber Journal* 1(7): 44–46.

Grant, R.K. (1979) The role of economics in forestry: the case for an independent view, *New Zealand Journal of Forestry* 24(1): 47–60.

Hargreaves, D. (1991) US giant eyes potential in the Pacific, *Evening Post,* 23 November.

Healy, B. (1981) *A hundred million pine trees*. Hodder and Stoughton, Auckland.

Jones, O. (1928) Joint stock forestry, *Empire Forestry Review* 7(1): 55–63.

Kirkland, A. (1981) *Future exotic forest development. First lead paper: needs and opportunities for planting*. New Zealand Forestry Conference, Wellington.

Kirkland, A. (1990) *The sale of Crown forest assets*. Address to the Forest Industries 1990 Conference, Rotorua.

Le Heron, R.B. (1988) The internationalisation of New Zealand forestry companies and the social reappraisal of New Zealand's exotic forest resources, *Environment and Planning A* 20: 489–515.

Le Heron, R.B. and Roche, M.M. (1985) Expanding exotic forestry and the extension of a competing use for rural land in New Zealand, *Journal of Rural Studies* 1(3): 211–29.

Meister, A.D. and Fowler, D.E. (1984) Planning limitations on forestry, *People and Planning* 31: 13–15.

Ministry of Forestry (1991a) *New Zealand Forestry Statistics 1991*. Ministry of Forestry, Wellington.

Ministry of Forestry (1991b) *Report of the Forestry Joint Venture Working Group*. Ministry of Forestry, Wellington.

New Zealand Forest Service (1983) *Man made forests: growing your future*. New Zealand Forest Service, Wellington.

Roche, M.M. (1991) 'Privatising the exotic forest estate: the New Zealand experience'. Paper presented to the History of the Forest Economy of the Pacific Basin Symposium, XVII Pacific Science Congress, Honolulu, Hawaii.

Smith, B. (1981) Rural change and forestry, *People and Planning* 20: 10–13.

State Services Commission (1982) *Report of the New Zealand Forest Service Review Committee*. State Services Commission, Wellington.
Thomson, A.P. (1975) *Land requirements and considerations for target achievement*. Forestry Development Conference, Wellington.
Valentine, J. (1989) Do we need a forestry policy?, *Forestry Forum* 4: 16.
Working Party on Afforestation (1981) *Afforestation Working Party Report*. New Zealand Forestry Conference.

# 11 Reafforestation in Australia

*G. T. McDonald*

## Introduction

Most of the Australian population lives in the forest biome – the more hospitable eastern and southern margins of the country. Only a small portion of Australia has the climate, soils and topography suitable for the growth of forests and woodlands. By 1990 only a tiny 5.3 per cent of the country was still forested. Throughout the European history of Australia, priority was given to agriculture and other more intensive uses rather than to forests, and a large portion of the original forests were cleared. Most rural uses need well watered, fertile and gently sloping land, just those lands with the best forests. Forest land is in most cases a residual: marginal land, too steep, too stony or too infertile, or too remote for agricultural activities.

Constitutionally, the states have the predominant power and responsibility for managing forest land resources under Australia's three-tiered system of government; at least it seemed that way until the early 1980s when the Commonwealth passed the World Heritage Properties legislation allowing the Commonwealth to influence forest land use through its powers over matters concerning international treaties, trade and the financial system. For over eighty years of federal government, land-use planning powers concerning forestry rested mostly with the states and while many day-to-day management issues are still resolved at the state level, there is an increasingly important role for the national government in forest policy in Australia.

Reafforestation in Australia takes two principle forms. Firstly, Australia has a substantial area of plantations, especially softwoods, grown for timber production. Secondly, revegetation for conservation reasons has resulted in a very large area of reafforestation on farms. Reasons for 'conservation' afforestation include goals such as reversal of tree decline, creation of wildlife habitats, prevention or reversal of land degradation, and shelter for crops, livestock and general amenity.

In past decades, the integration of commercial wood production, farming, conservation and natural resource management has not been formally acknowledged in the form of interlocking government natural resource policies from the sectors of agriculture, forestry, conservation, soil and water; hence, plantation programmes have been a more prominent feature of Australia's forestry history. They will be discussed first.

**Table 11.1** Land area by type of vegetation cover, 1987 (thousands of hectares)

| | NSW | Vic | Qld | WA | SA | Tas | NT | ACT | Total |
|---|---|---|---|---|---|---|---|---|---|
| Forest | 14,959 | 5,257 | 11,796 | 2,665 | – | 2,847 | 3,266 | 51 | 40,841 |
| | (18.7) | (23.1) | (6.8) | (1.1) | (0.0) | (42.0) | (2.4) | (21.3) | (5.3) |
| Woodland* | 3,300 | 3,010 | 28,200 | 20,530 | 900 | 1,051 | 7,000 | 5 | 63,996 |
| | (4.1) | (13.2) | (16.3) | (8.1) | (0.9) | (15.5) | (5.2) | (2.1) | (8.3) |
| Plantation | 230 | 211 | 185 | 87 | 91 | 83 | 4 | 14 | 905 |
| Total land area | 80,160 | 22,760 | 172,720 | 252,550 | 98,400 | 6,780 | 134,620 | 240 | 768,230 |

Figures in brackets are percentages of total land area.

*Includes woodland (946,000ha) which is vegetation dominated by trees with top height exceeding 5m, usually with a single stem, with tree crown cover in the range 10–30 per cent of the land area occupied; and scrubland (2,064,000 ha) which is vegetation with top height 2–8m containing shrubs, often multi-stemmed, and including mallee-type eucalypts.

*Source*: Resources Assessment Commission 1990.

## Forest and forestry in Australia

### *Australia's forests*

Australia's potentially exploitable productive commercial forests are situated in a discontinuous belt on the eastern margins of the continent, in Tasmania, and in the south-western corner, and have a total area of about 140 million ha. Economically exploitable remaining forested areas within this belt cover an area of 42 million ha, which is less than 6 per cent of Australia's total land area (Table 11.1). Most of these forests are broad-leaved (i.e. non-coniferous hardwood) types dominated by eucalypt species and in 1978, 77 per cent of all timber removed was eucalypt hardwood.

Private native forests comprise over 20 per cent of forested land, but their productivity is low and erratic and only half of that from state forests because of the poor original stocks and the lack of incentives for mixed agro-forestry uses. Relatively little private native forest is reserved for long-term wood production. Over 40 per cent of forested land is held in state control as vacant land or occupied under forms of lease that do not specifically secure this land for permanent timber production. Some of this land and some state forests may eventually be converted to other uses, especially national parks, but also agriculture, grazing, recreation, and for future urban development.

Indigenous species other than eucalypts, including rainforest species, are a small portion of the area and production of Australian forests. Plantation softwoods, mostly conifers, are already contributing over a quarter of all timber production, although they occupy about 2 per cent of the area of commercial forests. Softwoods are destined to contribute about 60 per cent of total wood production early in the next century. Most of these plantations are exotic pine, especially *Pinus radiata* in southern locations and *Pinus elliottii* and *Pinus caribaea* in Queensland. Only a small area of native hardwoods and softwoods are grown under plantation conditions in Australia (see Table 11.3).

There are also non-utilitarian values to be considered in forest management, including the conservation of natural areas in their own right, for scientific and educational purposes or for their aesthetic values. Many of these contributions of forests to social well-being and community cultural needs are met in the system of national parks, but there are some who feel that the nation already has too little natural forest in a general sense, and that there are areas of forest that should be withdrawn from logging and kept for nature conservation purposes (Dunphy 1978; Mosley 1981).

### *Brief history of forest exploitation*

Carron (1979) summarises the main phases in the evolution of forest planning in Australia since European settlement:

(i) The pioneer settlement phase of the latter part of the nineteenth century saw massive clearing of forest land for agriculture, and there was exploitative logging to meet local needs and earn some export income for high-valued timber such as cedar. The main function of forest administration at this time was to collect royalties and administer exploitation programmes. Forest management was a part of general land management (see also Carron 1986).

(ii) During the 1870s, there was some public and official recognition that destruction must be curbed and forests seen as resources in their own right, capable of earning revenue for government and not just obstacles to agricultural development.

(iii) During the 1920s, government organisations dealing with forestry began to adopt their current role of securing land for forestry purposes and maintaining a stable supply of timber for construction purposes. This broadening of the scope of government forestry took place in the face of great pressure by other organisations seeking land for agricultural settlements.

(iv) During World War II, Australia was forced to increase local timber production to offset the loss of imports, and became aware of its shortage of forest products. Consequently, by the early 1970s, all states had begun plantation programmes which though small in area were disproportionately important in their contribution to output (and to perceived environmental effects). These expanded markedly with the stimulus of federal government finance in the 1960s and 1970s. There was growing sophistication in forest management but 'much of Australian forestry was still in a transition between the pioneering problems of exploiting a wild, if much modified, forest, and the modern ones of managing successive crops for sustained yields in an environmentally harmonious manner' (Dargavel 1987, p. 7).

(v) The period of the conservationist area and the involvement of the Commonwealth in forest management, which commenced with the passage of the Australian Heritage and World Heritage Acts in the late 1970s and early 1980s, has only just begun, and planning policy and practice are still in a transitional phase.

**Table 11.2** Production, consumption and trade in sawn wood, classified by type, 1986–87 (cubic metres)

| | | | |
|---|---|---|---|
| Production | | | |
| Plantation conifers | | 1,100,659 | |
| Native forest timbers | | | |
| Cypress pine | 92,815 | | |
| Other conifers | 26,180 | | |
| Eucalypts | 1,669,791 | | |
| Other broadleaved* | 38,257 | | |
| | | 1,827,043 | |
| Total production | | | 2,927,702 |
| Imports | | | |
| Broadleaved | 230,372 | | |
| Coniferous | 836,794 | | |
| Total | | 1,067,166 | |
| Exports | | | |
| Broadleaved | 21,676 | | |
| Coniferous | 830 | | |
| Total | | 22,506 | |
| Consumption | | | 3,972,372 |

*Includes brushwood, cabinet timbers and other broadleaved species
*Source*: ABARE (1988)

## Forest planning

### *Overview*

As a result of the decentralised administration of Australian forestry amongst the states, Australia does not have what could be termed a national forest policy, only a record of actions at the national level, and an aggregate of state actions in managing their reserves – an implicit forest policy.

These actions have been taken at the national level with respect to trade, softwood agreements, environmental controls on pulp mills, heritage legislation, impact assessment and related administrative decisions. In aggregate these actions are *ad hoc*. Such *ad hoc* policy is not unusual in Australia due to division of powers but increasingly the limitations of such approaches are becoming clear, especially as far as the balance between the use of forests for wood production and their use for recreation and conservation is concerned.

Australia has been committed since the national FORWOOD Conference of 1974 to a policy of net self-sufficiency in forest products. That is, it aims to have a net trade balance in forest products. As shown in Table 11.2, Australia imported one-quarter of its sawn-wood requirements in 1987. It also imports an even larger portion of its requirements of paper and pulp. A commitment to self-sufficiency was reflected in the Commonwealth Softwood Forestry Agreement Acts, as amended (1967–76), under which Commonwealth funding of state softwood programmes was granted. The various state forestry agencies also aim to achieve state self-sufficiency in the long term, an important element in their decision-making.

Without far-reaching trade restrictions or demand management, the only way that Australia could achieve (net) self-sufficiency would be to subsidise the supply of domestic wood. As a means of reaching self-sufficiency, softwood plantations, whatever their economic worth, were chosen as the way to do so. From a land-use-planning point of view, the principle of self-sufficiency, national, state or regional, has had a number of important consequences:

(i) It has provided a basis on which the long-term development of wood-based industries can proceed.
(ii) It has been the driving force behind rapid expansion in the area of exotic softwood plantations in all states.
(iii) It has been used as the foundation of other forest planning activities, providing a guideline for determining forestry land requirements and a defence of forestry claims for land for which there are competing interests. It has provided options for the management of native forests which are increasingly the object of conservation claims or of declining productivity.

### *The plantation programme*

The areas of plantation forest in Australia are shown on Table 11.3. While the areas are small in comparison with the native hardwood forests, the productivity is relatively high, and industry increasingly relies on them for timber and pulp materials. The plantation forests can be described on a regional basis (see also Figure 11.1):

(i) Southern NSW plantations, almost exclusively of *Pinus radiata* grown in state forests in the Bathurst–Tumut region of the southern Highlands. The plantations are the basis of large integrated forest products industries in the region including saw mills, processed board mills and pulp mills.
(ii) South-east Queensland, in the Gympie–Maryborough area. Here plantations of mostly *Pinus elliottii* grown in state forests on coastal lowlands (wallum lands). Private plantation forests in the Caboolture district closer to Brisbane established by AMCOR (then APM) have been sold to a land development company (See Brown *et al.* 1988).
(iii) The *Pinus radiata* plantations of north-east Victoria.
(iv) The *Pinus radiata* and indigenous hardwood plantations of Gippsland.
(v) The *radiata* plantations of south-eastern South Australia. In South Australia the native forest resources are very limited due to climatic constraints and there has been a long history of plantation forestry in that state. The state-owned forests and private forests are in the Mount Gambier region where they form the basis of a saw-milling, ply mills, pulp mills, particle board and preservation plants.
(vi) Western Australia plantations are in the Blackwood River valley, near

**Table 11.3** Plantation area by species for Australia and the states, 1987 (ha)

| State | Coniferous species | | | | | | Broadleaved species | | |
|---|---|---|---|---|---|---|---|---|---|
| | *P. rad* | *P. elli* | *P. pina* | *P. carr* | *Arauc* | Other | *Euc* | *Pop* | Other |
| NSW | 212,121 | 5,186 | 0 | 2,786 | 1,582 | 5,231 | 1,202 | 1,877 | 0 |
| Vic | 192,473 | 0 | 0 | 0 | 0 | 4,330 | 13,614 | 258 | 71 |
| Qld | 3,227 | 90,206 | 0 | 40,191 | 44,401 | 5,229 | 1,328 | 0 | 458 |
| WA | 48,934 | 269 | 28,012 | 0 | 0 | 420 | 9,470 | 0 | 0 |
| SA | 85,961 | 0 | 3,434 | 0 | 0 | 374 | 1,199 | 0 | 0 |
| Tas | 64,967 | 0 | 0 | 0 | 0 | 333 | 15,197 | 0 | 2,663 |
| NT | 0 | 0 | 0 | 2,400 | 0 | 1,801 | 0 | 0 | 0 |
| ACT | 13,365 | 0 | 0 | 0 | 0 | 514 | 0 | 0 | 0 |
| Total | 621,048 | 95,661 | 31,446 | 45,377 | 45,983 | 18,322 | 42,010 | 2,135 | 3,192 |

*rad: radiata; ell: elliottii; pin: pinaster; carr: carribea; Arauc: Araucaria; Euc: Eucalyptus; Pop: Populus.*

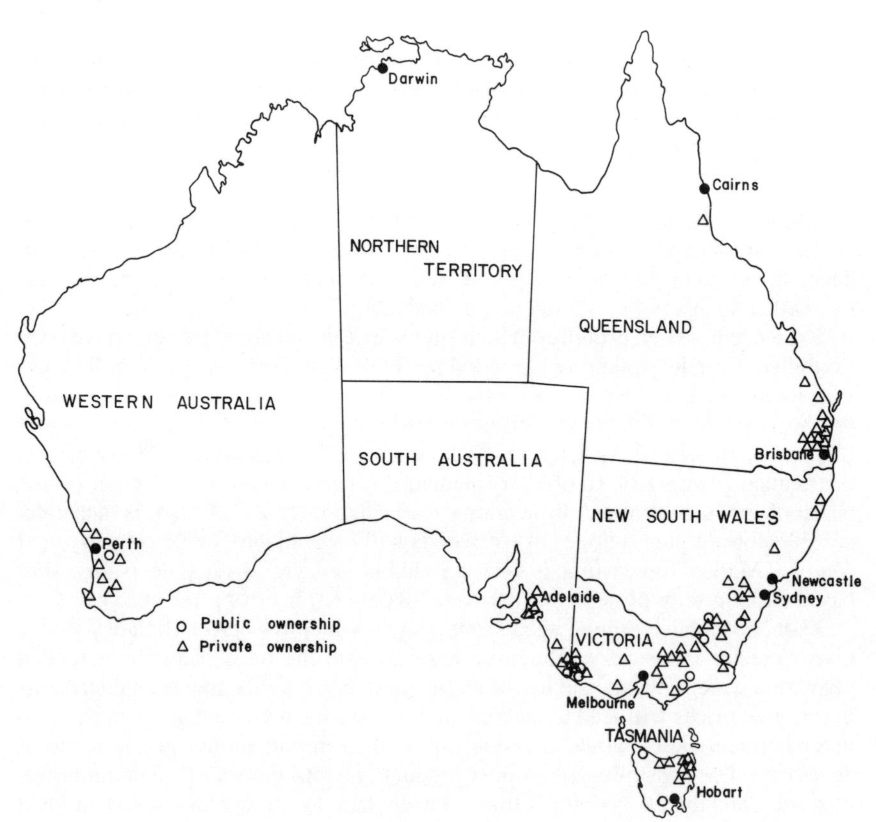


**Figure 11.1** Distribution of state and private plantations (areas over 2,000ha).
*Source*: after Booth (1984)

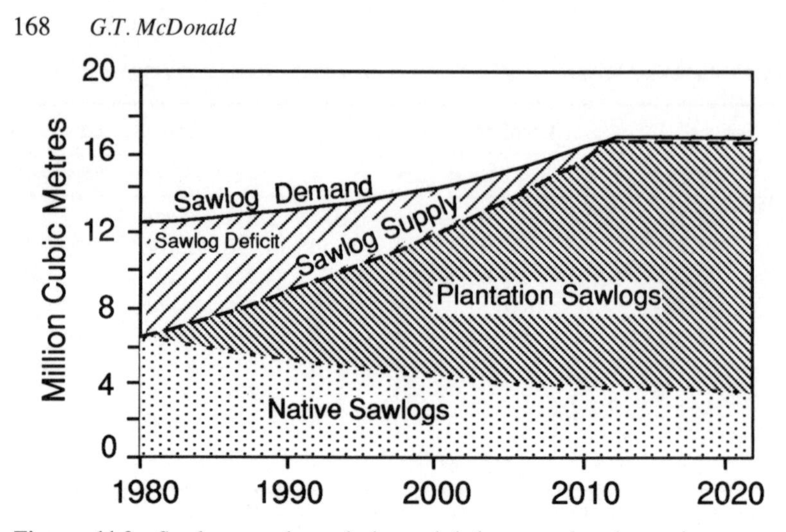


**Figure 11.2** Sawlog supply and demand balance under the regime proposed by FORWOOD (1975).

Nannup north of Perth and the Donnybrook Sunkland district between Collie and Busselton where plantations are located in large forest reserves. The Western Australia plantations consist of *Pinus radiata* and *Pinus pinaster*, the latter being more tolerant of poor soils.

(vii) Tasmanian *Pinus radiata* and eucalypt plantations.

The self-sufficiency policy has been criticised on a number of grounds. Some of these are technical, such as disagreement on the level of future demand and hence the scale of plantation required, and some more fundamental criticisms as to whether such a policy should be pursued at all.

Evidence has been produced which shows that the levels of planting have been excessive. Certainly planting exceeded the FORWOOD targets of 28,500 ha per year by as much as 25 per cent in the late 1970s, and surpluses would be likely on the basis of the demand estimates made at that time. Government reports (Australia, Bureau of Agricultural Economics 1977; Treadwell 1978) suggested that annual planting of 16,665 ha (median estimate) would be sufficient on the basis of revised demand projections (see Figure 11.2). If that is accurate, considerable surplus supplies of softwoods will be available for export early next century. Market forecasting is very unreliable and yet forestry decisions were based on long-term projections that were 30 per cent in error after five years!

Economists have argued against the way in which the self-sufficiency policy gives forestry industries preferential assistance in the form of publicly funded plantation developments and use of public land. It is claimed that such distortions in resource prices will lead to inefficiency in resource use and that the nation and its constituent states would be better advised to import timber requirements if necessary. The possibility of surplus production from the states' plantation programmes, and the likely competition between them to attract major forest-product industries, could lead to lower public returns from the forests and a transfer of this surplus to the producers. It also gives rise to the likelihood of Australia exporting some forest products from its plantation forests. Import competition will become increasingly significant as Southern Hemisphere plantations achieve large export surpluses. Competition may be expected especially from New Zealand, which benefits from the Closer Economic Relations treaty ensuring open access to Australian markets.

From an environmental point of view, the plantation policy has had a mixed reception, although on balance it has the support of the Australian Conservation Foundation (ACF). Environmental impacts are minimised when the plantations are established on already cleared land and not at the expense of low (wood) value native forests as has been the case until recently. Even though it may not agree with all the details of the plantation program, the ACF accepts that it has the major environmental advantage of lessening the pressure on other forest resources, including rainforest, especially within Australia but also in South-East Asia. The natural forests saved may have higher conservation values than the land used for plantations, and an important issue in assessing the environmental impacts of the plantation programme is the choice of sites. Conservation interests favour the use of degraded farmland for plantation purposes.

At the same time, critics have asserted that there are adverse environmental effects of conifer plantation development. These include the facts that:

(i) natural forest is sometimes destroyed without adequate concern for its intrinsic values, such as wildlife or landscape;
(ii) soils may be degraded by podsolization and lower accumulation of organic matter;
(iii) land clearing may cause accelerated erosion and siltation and have adverse hydrological effects in catchment areas;
(iv) the use of chemicals involves risks to the plantation and adjoining land; and
(v) noxious weeds may become more prevalent and difficult to control.

Exotic pine plantations have been labelled as biological deserts by some conservationists. Although this point of view has been shown to be extreme, since some species of native fauna adapt to these artificial habitats, there is of necessity some loss of natural habitat. This should, however, be seen in context of a target of 1.14 million ha of plantation forest compared with an area of about 20 million ha of agricultural cultivation in Australia.

Single-species plantations over large areas may risk disastrous outbreaks of pests and diseases not likely in multi-species forests. Wood wasps and needle blight have affected *Pinus radiata*, but fortunately so far there have been no serious outbreaks. Fire is also a major hazard, as was discovered in South Australia in 1983 when a significant portion of that state's *Pinus radiata* plantations was burned (see Keeves and Douglas 1983). It is possible that careful design of plantation layouts can minimise the ecological impact of single-species plantations, and research by Friend (1980) has shown that leaving natural-habitat refuges can substantially improve the survival chances of both flora and fauna species. Moreover, by ensuring a better ecological balance, greater numbers of birds will serve as predators for undesirable insects.

Rural region economic development has been a major objective of forestry activities in all Australian states. To evaluate the economics of this argument, all costs should be counted, including intangible environmental costs, and a comparison made between forestry and other potential users of the same resources such as agriculture. Given that forestry will supply mainly relatively secure domestic markets, it should provide an important stabilising and diversifying component of the economic base of a number of depressed rural regions. To the extent that decentralised population and industry are seen as socially desirable, plantation forestry is beneficial. Others have argued that forestry has not been an economic use of national resources and that the economic benefits are overstated (Dargavel 1982; Cameron and Penna 1988).

Conservationists argue that logging of native forests should be phased out over the next 10–15 years and be replaced by protection status for other uses. Any logging over the intervening period should be restricted to small volumes of high-value timber on a very long rotation. The ACF policy calls for plantations of both hardwood and softwood to provide most of the country's wood requirements. 'The plantation, like the farm and the orchard, offers tremendously increased production in return for greater initial investment and concentrates the disturbance of the environment in a much smaller area, leaving a much larger area free to satisfy other values' (Wootten 1987, p 70).

There has been little incentive to establish eucalypt plantations in Australia when such wood was freely available from natural forests. The same could be said about the high-valued cabinet timbers found in the country's rainforests.

## Regreening programmes – conservation tree planting

Reafforestation, or 'greening', is a recent major dimension of conservation activities in Australia. A great variety of business, government and community organisations at national, state and local levels are directly or indirectly involved in the timber or agricultural sectors, or in the management of natural resources, including revegetation: schools, municipalities, farmers, community groups, plantation owners, and state (National Plantations Advisory Committee 1991, p 269). The public and private bodies provide assistance in the form of free seedlings, grants, technical advice on tree-planting techniques, signs, and fencing.

Regreening for conservation purposes in Australia operates in an interlocking manner with involvement from all levels of government as well as non-government organisations and private companies that sponsor tree-planting activities. Constitutionally, the federal government can only fund and assist afforestation schemes for conservation purposes. All federal programmes call for community awareness and initiatives to address soil, water and vegetation concerns at a district level. As soil, water and vegetation are interrelated, interdependent elements of the land, programmes recently established by the federal government to address natural resource problems such as land degradation, salinity, and tree decline all have overlapping causes and activities. A network of afforestation programmes for conservation has only emerged in the 1980s with the National Tree Program (NTP). More recently, the One Billion Trees Program has taken the place of NTP and emerged as the main afforestation assistance programme for conservation purposes. The other major national reafforestation programmes include the National Soil Conservation Program, the Murray–Darling National Afforestation Plantation programme, Greening Australia, and Save the Bush.

Local governments involved in both state and federal land-care initiatives have the infrastructure for community involvement which can be used to lead the community to higher levels of awareness and responsiveness to implement programmes that address individual community land-care needs.

State government agencies complement and develop further federal initiatives. 'State Governments have also developed totally independent government initiatives to foster integrated resource management' (National Plantations Advisory Committee 1991, p 270). When considering the practicalities of implementing regreening projects, all tree-planting programmes, regardless of their importance in afforestation policy, require the following considerations for their ultimate success:

(i) What are the conservation objectives:
 (a) tree planting for conservation;
 (b) creation of wildlife habitats;
 (c) the prevention or reversal of land degradation;
 (d) shelter for crops, livestock and general amenity.
(ii) Selection of species, taking into account biological, terrain and environmental aspects of the site and the available tree species and seed collection.
(iii) Suitable revegetation methods:
 (a) natural regeneration;
 (b) direct seeding (time of seeding);
 (c) planting seedlings (time of planting).
(iv) Site preparation techniques:
 (a) weed control;
 (b) protection from grazing.
(v) Maintenance:
 (a) fire management;
 (b) pest control;
 (c) fertilising.
(vi) Availability of labour, volunteer or paid, to carry out each of the above tasks (Venning 1988, pp vii, viii).

These technical, planning and administrative requirements are addressed in many afforestation schemes. Most assistance schemes offer financial aid in some form or another. Farmers, conservation groups, schools, community groups and other interested parties can choose from a variety of assistance programmes. Each scheme has a particular aim, whether it be prevention of soil degradation, protection of wildlife habitats, reversal of tree decline or to provide a source of timber. Schemes are targeted at particular groups such as landowners, schools, municipalities or commercial foresters.

*Commonwealth afforestation policy and planning*

As afforestation policy and planning is largely the responsibility of the individual state governments and their respective forestry agencies, the federal government has a limited role in the day-to-day management of forestry issues and in balancing the multiple uses of forests, mainly being conservation, logging, catchment protection, and recreation. State forestry agencies act not only as protectors of state-owned forests but also as commercial organisations, in their own right, generating revenues from the 'wood-value' of forests and fulfilling state revenue requirements. The federal government, on the other hand, has taken the role of financial sponsor and educative and promotional facilitator of afforestation programmes targeted at community groups, private individuals and landholders with the aim of increasing afforestation activities throughout the country. The federally assisted tree planting programmes should be seen as secondary but complementary to the afforestation policies of each of the state forestry agencies.

The afforestation assistance schemes for conservation began in the early 1980s with the initiation of the National Tree Program (NTP). There have been a succession of similar strategies to complement the large number of small local conservation groups involved in tree establishment. However, for the purpose of providing an account of Australia's experience of afforestation, only the NTP will be discussed.

In 1982 the NTP was established as a catalytic programme for future tree-activity programmes. The main aim of the NTP, endorsed by the Minister of Arts, Heritage and Environment in 1985, was to conserve and establish trees and associated vegetation for community and private benefits throughout Australia. Its objectives include increasing the tree cover in selected rural areas; promoting co-ordinated action by individuals, government and the community generally to conserve, plant and regenerate trees; and developing public awareness of the value of the tree (Dept. of the Arts, Sport, the Environment, Tourism and Territories (DASETT) 1988). These objectives became the basis on which a succession of tree-planting schemes were drafted, especially the One Billion Trees Program.

The NTP was administered through the National Co-ordination Committee (NCC), comprising representatives from each of the states and territories, local governments and a range of non-government organisations such as the Australian Conservation Foundation, National Farmers' Federation, Australian Council of Local Government Associations, Greening Australia, Institute of Foresters of Australia and Men of Trees.

One of the most ambitious initiatives in the implementation of tree planting trees on a broad scale was the establishment of a formal link between NTP and the Community Employment Program (CEP). The CEP was designed to create employment opportunities for the unemployed. Its funds were directed to creating employment in demonstration projects, tree-planting activities and research. By creating CEP secretariats in the states and territories, applications for funding of tree-planting projects are referred to the corresponding NTP Co-ordinating Committee (DASETT 1988).

Between August 1983 and August 1986, $24 million was expended on 600 tree-related projects, employing over 4200 people. The majority of the funds were allocated to forest production/recreation (29 per cent), rural revegetation (25 per cent), and urban amenity/landscaping (25 per cent). Financial assistance was also given to rural amenity/national parks, research information, and nursery development (DASETT 1988).

Many positive things have resulted from this initiative, notably the employment and work experience to participants of the programme, the increase in the number of trees, and the public awareness and exposure that the NTP received. However, there were a number of limitations to the application of the CEP to tree projects. Employment was determined by the seasonal nature of tree establishment and the maximum period of employment being six months. In addition, skilled workers were needed to strengthen the training and guidance of unskilled workers.

Despite these problems, the experience of NTP provided benefits to future schemes as a source of knowledge of the limits and strengths of the implementation of community-based afforestation programmes. Lessons have been learnt. The One Billion Trees Program, the successor of the NTP, relies mostly on demonstration programmes and field days to teach applicants about the most successful way of revegetating land, rather than on employment schemes.

### *Current Commonwealth afforestation initiatives*

In response to the growing public concern about land degradation in Australia, the Commonwealth government has developed the National Land Care Program, first announced on 20 July 1989, in Prime Minister Hawke's launch of a Statement on the Environment. The Commonwealth government's afforestation policy and planning is addressed as a component of the National Land Care Program under the One Billion Trees Program (OBTP).

The Land Care Program is more formally described as a broad range of national assistance programmes offering funding and education for activities dealing with the management of the land, water, trees and vegetation resources of Australia according to the Commonwealth government publication *Our Country, Our Future* (July 1989). Although the Land Care Program is a federal government initiative, the involvement of the state and territory Land Care groups, local governments, landholders, private individuals and the community at large is necessary for the success of the overall operation of OBTP. The federal government acts as a coercive and promotional facilitator providing both education and financial assistance to state and territory Land Care groups.

In addition to initiating the Land Care Program, Prime Minister Hawke declared 1990 as the Year of Land Care and the 1990s as the Decade of Land Care. For the next decade, the Land Care Program will play an educative and sponsoring role in protecting, managing and repairing the environment. Government funding to be distributed over the Decade of Land Care includes approximately $320 million to the National Soil Conservation Program (NSCP), $49 million to the OBTP and $20 million to the Save The Bush Program.

Although all programmes under the National Land Care Program include afforestation as one of their activities, afforestation is most directly addressed under the OBTP: 'One Billion Trees is a program for the re-greening of Australia. It provides for the planting of one billion trees by the year 2000' (Hawke 1989, p iv). The OBTP centres on community involvement and does not assist state governments in their respective forest-management activities, but gives assistance to local governments as well as to private individuals and small Land Care groups.

The One Billion Tree Program replaced the activities of the earlier (1982–87) National Tree Program. The main goal of both the OBTP and NTP was the reversal of tree decline in Australia. The OBTP's objective of planting 1 billion trees by the year 2000 will be achieved by increasing public awareness of the value of trees and promoting action at individual, government and community levels to plant trees.

In *Our Country, Our Future* OBTP is defined as:

(i) A Community Tree Planting Program to plant over 400 million trees. This will include the following:
(a) financial assistance for community groups and landholders to implement tree projects on farms and in towns and cities;
(b) a schools nursery project to provide a hands-on learning experience for young people; and
(c) major projects involving participation by community, corporate and government organisations.

(ii) A Natural Regeneration and Direct Seeding Program to establish over 600 million trees in open areas of Australia commencing with trials and demonstrations across the country to improve methods of growing trees and to encourage wide-scale action by landholders (Hawke 1989).

*Greening Australia – contractor and promotional facilitator of the OBTP*

The One Billion Trees Program is administered by Greening Australia, a national community, non-profit, non-government organisation. Greening Australia first emerged in 1982 as what was at first a central organising committee undertaking tree projects and promotional activities for the United Nations Association of Australia 'Year of the Tree'. It has continued and grown through increased membership to become a loose yet cohesive umbrella for a great diversity of individuals, groups and institutions with an active interest in overcoming tree decline and land degradation, particularly in rural areas (DASETT 1988).

Greening Australia branches are located in all of the state and territory capitals as well as in most cities and larger rural centres. As Greening Australia is a network of well-established branches of community-based organisations, it is the most suitable contractor and promotional facilitator existing in Australia to achieve the objectives of the OBTP. The latter, a representative of community afforestation planning and practice, seeks to complement the state government revegetation policies in areas where state governments have little or no control (e.g. privately owned farms, urban areas and municipalities).

As part of the instigation of OBTP, Greening Australia branches in each state and territory capital have been assigned project officers trained in the field of forestry. State and Territory Committees of Greening Australia distribute information, endorse tree activities and projects, raise funds and provide a link between government and non-government organisations (DASETT 1988).

Building from these original activities of Greening Australia, OBTP has developed the above-mentioned goals in the following way:

(i) Liaising with the pre-existing infrastructure of forestry officers from state government departments and local Land Care groups to distribute information about revegetation techniques. The pre-existing network of Land Care officers acts as an intermediate link between Greening Australia, representing OBTP objectives, and clients such as individuals, landholders, local groups, etc. Forestry officers or project officers can provide technical tree-planting advice on such matters as species selection, revegetation method, site plan, etc.

(ii) Assessing applications of clients for funding for tree-planting projects. Selection criteria for receiving funding and assistance for tree-planting projects varies across each of the state and territory branches. For example, in Victoria and South Australia, most applicants are local Land Care groups made up of individuals owning properties linked by physical proximity and terrain attributes such as water-catchment boundaries. In Queensland, most applicants are individual private farmers as it is difficult for farmers to establish groups when such large distances exist between neighbouring farms. In some states, assistance is only given to individuals or groups that have a specific land-degradation problem on their property. In other states, applicants may be local-government parks departments requiring novelty trees such as Jacarandas or local primary schools interested in setting up nurseries and shadehouses for rainforest species.

(iii) Providing funding and seedlings or information about where seedlings can be obtained from. Often school groups that have established nurseries are sources of free seedlings for applicants of tree-planting projects.

(iv) Demonstrating natural regeneration techniques and planting of new seedlings in environmental-awareness days and field days.

(v) Promoting the philosophy behind Greening Australia and the One Billion Tree Program through means such as newsletters, pamphlets, and school visits.

(vi) Raising funds through membership, selling of products with the Greening Australia logo, and recruiting sponsors such as Alcoa in Western Australia and other large companies.

### *Other national afforestation schemes*

Although OBTP addresses afforestation most directly, there are many other national programmes, both government and non-government, which deal more generally with natural resource management (ie. soil, water and vegetation) but which incorporate afforestation as one of their objectives. Other major national programmes include the National Soil Conservation Program (NSCP), the Murray-Darling Basin Natural Resources Management Strategy Program, Save the Bush Grant Scheme, Aboriginal Employment Development Program, and the Australian Trust for Conservation Volunteers.

#### *The National Soil Conservation Program*

The NSCP is administered by the Land Branch of the Federal Department of Primary Industries and Energy.

> This program has had a major impact on the development of landcare and tree farm groups around the country. It concentrates on land degradation controls and development and adoption of sustainable land use practices, with provision for funding of

landholder based landcare groups, State agencies, Local Government and other community groups (National Plantations Advisory Committee 1991, p 275).

The NSCP only funds tree-planting projects whose primary purpose is to demonstrate treatment of the land degradation problems. An assessment panel approves applications. Land Care groups and other community groups are eligible for receiving assistance with the restriction that Land Care groups must contribute 50 per cent of the cost in cash (Boutland *et al.* 1990). Grants of up to $15,000 per annum are available to State-approved landholder-based community projects for up to three years.

*The Murray–Darling Basin Natural Resources Strategy Program*

The Murray–Darling Basin Natural Resources Strategy Program is administered by the Murray–Darling Basin Commission. Its purpose is to 'develop integrated and sustainable development of natural resources (land, water, environment and cultural) of the Murray–Darling Basin through on-ground community activities, which are supported by the government' (Boutland *et al.* 1990, p 23). The aims of this strategy include, amongst others, preventing further degradation of natural resources, restoring degraded resources and promoting sustainable use practices (National Plantation Advisory Committee, 1991). These aims relate directly to revegetation and so technical, administrative and financial assistance for tree-planting activities is offered in the form of planning and implementation of on-ground activities, knowledge and advice about tree planting and vegetation retention and management, and community education projects (Boutland *et al.* 1990).

*The Save the Bush Grant Scheme*

The Save the Bush Grant Scheme is administered by the Australian Parks and Wildlife Service. Its purpose is to 'protect and manage remnant native vegetation, particularly in the context of maintenance of biological diversity' (Boutland *et al.* 1990, p 24). Eligible applicants include community, farming and environmental groups, and other relevant organisations, such as schools, academic research organisations, to state, territory and local government departments and agencies. Financial assistance is available for on-ground vegetation protection activities, demonstrations of bush-conservation practices and activities, research, and data collection, management and dissemination. This programme, however, is unlikely to provide grants for tree planting. This scheme is worth acknowledging because it can be involved in the dissemination of knowledge about native species and other conservation practices important to the technical implementation of tree planting.

*The Aboriginal Employment Development Program*

The purpose of the Aboriginal Employment Development Program is to 'create employment and enterprises for Aboriginal people living in rural and remote areas through land management projects, specifically pest animal and plant control, and wildlife utilisation' (Boutland *et al.* 1990, p 26). Assistance is in the form of funding, scientific and technical advice. Aboriginal people must be employed. Tree planting is included as a rehabilitation component of pest control projects.

*The Australian Trust Conservation Volunteers*

The Australian Trust Conservation Volunteers (ATCV) is a large non-government organisation which operates from Ballarat, Victoria. It is a non-profit, non-government, national scheme that is funded by the Commonwealth government. It provides a link between landholders with conservation problems and volunteers willing to help with the necessary work. It offers assistance in the following form:

(i) labour to plant trees, eradicate rabbits, build fences, collect seed, etc.;
(ii) a trained team leader, up to 10 volunteers, a vehicle and basic hand tools are provided;
(iii) ATCV acts as a management organisation for volunteers to work with landholders.

Each of the states and the Northern Territory have their own independent programmes and branches of government departments addressing land-care issues and afforestation. Government and non-government, federal, state and local schemes and organisations interlock creating a intricate network.

*State afforestation schemes – a Victorian case study*

This discussion above highlights the fact that community based reafforestation is a major national priority in Australia and that there are many well-funded programmes in place. To illustrate the operation of these programmes, the example of a Victorian farmer may be considered.

Such a farmer wishing to conserve some remnant vegetation in order to conserve wildlife as well as to plant some trees for a timberbelt can turn to a number of assistance programmes. They include the following:

(i) The Murray–Darling Basin Natural Resources Management Strategy Program offers funds for planning and implementing tree planting or vegetation management to prevent land degradation.
(ii) The Australian Trust for Conservation Volunteers may be able to provide additional labour for tree planting and conservation works.
(iii) The Land Protection Scheme offers grants for planting softwoods and hardwoods.
(iv) Greening Australia may provide advice and limited assistance with fencing materials or costs.

The Land Protection Incentive Scheme is managed by Victoria's State Department of Conservation and Environment. It offers landowners planning and advice as well as financial assistance in the form of grants for softwood planting (up to $300 per hectare), hardwood planting (up to $600 per hectare), and agroforests (up to $300per hectare). This is only one of the many permutations of federal, state, and community-based projects that can be created to suit the objectives of the tree-planting project, the client and the site location.

*Evaluation of the assistance programme for conservation tree planting*

The major criticism of national tree-planting schemes for conservation and of the Commonwealth afforestation policy over the years is that the approach to revegetation remains somewhat haphazard and unsystematic. The *ad hoc* nature of these programmes reflects what sociologists and politicians describe as healthy community participation and involvement. The One Billion Trees Program serves as a mechanism from which the Australian 'community' at large can actively play a role in reversing tree decline. However, 1 billion trees alone will not make a substantial impact on achieving all conservation objectives. According to the Commonwealth Science and Industry Research Organisation, 20 billion trees would be required to alleviate significantly the salinity problem within the Murray–Darling Basin alone, which is only a fraction of the land needing revegetation. Hence, another criticism of the federally-assisted afforestation programmes is, in the short run, that their significance in the achievement of the goal of reversing tree decline in Australia is relatively small compared to the ongoing logging and clearing activities.

## Conclusion

Through just over 200 years of European settlement, Australia's forests were cleared to make way for agriculture and grazing, and productive forests were over-exploited. The costs of this mismanagement are now obvious. Economically, insufficient supplies of local timber and wood products result in substantial import requirements. Environmentally, the loss of trees causes land degradation through soil erosion and salinisation. Water quality has declined, flooding has increased and water supplies are less reliable through catchment degradation. Afforestation is now a major priority for governments at all levels, for farmers and timber companies and for the community as a whole.

Commercial plantations owned by private companies and by public agencies have expanded enormously over the past two decades, to the point where self-sufficiency and even net exports of wood products are likely in the next decade. Tree farming is becoming the predominant source of timber for the Australian market, and natural forests are becoming of declining importance in this respect. The natural forest is regarded more as a conservation resource. Regreening, reversing the decimation of trees in rural Australia, generally is now a well-established community goal. Governments are supporting these activities with substantial finance and with redirected technical activities. The enthusiasm to regreen the country has produced a plethora of programmes, perhaps not well co-ordinated, but they are wholly community-based. Changing attitudes and achieving responsible land stewardship depends almost entirely on such community involvement and commitment.

There is still a wide gap between the environmental needs for tree planting and for tree protection. One Billion Trees is an appealing slogan but the number needed is an order of magnitude greater at least. In concentrating on reafforestation, this chapter has emphasised the replanting of trees where trees once grew. This is important, but the overall management of forests and woodlands is an even more important issue in Australia. Natural regeneration of forests and woodlands is a much more efficient means of achieving a greener Australia. This depends on more appropriate catchment management and use of fire, and control of grazing animals and pests. Trees will regrow in the wetter parts of the country if given a chance. Unfortunately, while resources are being expended on tree planting in some districts, substantial areas are still being cleared in other districts for agricultural and grazing purposes. Control of this activity is a high priority. But that is another story.

## References


ABARE (1988) *Australian forest resources*. AGPS, Canberra.

Australia, Bureau of Agricultural Economics (1977) *The Australian softwood products industry*. AGPS, Canberra.

Booth, T.H. (1984) Major forest plantations in Australia: their location, species composition and size, *Australian Forestry* 47: 184–93.

Boutland, A., Byron, N. and Prinsley, R. (1990) *1991 Directory of assistance schemes for trees on farms and rural vegetation*. Greening Australia Ltd, Canberra.

Brown, A. L., Hindmarsh, R. J., McDonald, G. T. and Stock, E. C. (1988) Where have all the forests gone: forest land conversion in south-east Queensland, *Urban Policy and Research* 6(2): 51–64.

Cameron, J. I. and Penna, I. W. (1988) *The wood and the trees*. Australian Conservation Foundation, Hawthorn.

Carron, L. T. (1979) Forestry in the Australian environment – the background, *Australian Forestry* 42: 63–73.

Carron, L. T. (1986) *The history of forestry in Australia*. ANU Press, Canberra.

Dargavel, J. (1982) Employment and production: the declining forest sector re-examined, *Australian Forestry* 45: 251–61.

Dargavel, J. (1987) 'Whose prospects?', in J. Dargavel and J. Sheldon (eds.) *Prospects for Australian hardwood forests*. ANU CRES Monograph 19, Canberra.

DASETT (Department of the Arts, Sport, the Environment, Tourism and Territories) (1988) *National tree program progress report 1982–1987*. AGPS, Canberra.

Dunphy, M. (1978) The deforestation of Australia, *Habitat* 7: 14–15.

FORWOOD (1975) *Forestry and wood based industries development conference*. AGPS, Canberra.

Friend, G. R. (1980) Wildlife conservation and softwood forestry in Australia, *Australian Forestry* 43: 217–24.

Hawke, R.J. (1989) *Our country, our future*. AGPS, Canberra.

Keeves, A. and Douglas, D. R. (1983) Forest fires in South Australia on 16 February 1983 and consequent future forest management aims, *Australian Forestry* 43: 148–62.

Mosley, G. (1981) Why we need a national reafforestation plan, *Habitat* 9(2): 22–25.

National Plantations Advisory Committee (1991) 'Appendix C4: Community involvement in plantation development, existing mechanisms and information. Working Group 3. Farm planning and community involvement,' in *Integrating forestry and farming: commercial wood production on cleared agricultural land*. Department of Primary Industries and Energy, Canberra.

Resources Assessment Commission (1990) *Australia's forest and timber Resources*.

Resources Assessment Commission, Canberra.

Treadwell, R. F. (1978) Australian softwood plantation requirements, *Quarterly Review of Agricultural Economics* 31: 28–50.

Venning, J. (1988) *Growing trees for farms, parks and roadsides*. Lothian Publishing Company Pty. Ltd, Melbourne.

Watt, A. J. (1976) The policy process in the resolution of land use conflicts on the Boyd plateau, *Public Administration* 35: 212–28.

Wootten, H. (1987) 'Conservation objectives,' in J. Dargavel and J. Sheldon (eds.) *Prospects for Australian hardwood forests*. ANU CRES Monograph 19, Canberra.

# 12 Modern development of afforestation in Japan: process and results

*Yoshihisa Fujita*

## Introduction

About 70 per cent of the land area in Japan is mountainous. Most of the mountain area is under forest, while most of the land in the lowlands is in arable use (Figure 12.1). More than 40 per cent of the forest area consists of forests that have been planted in the last hundred years. The area of planted forest at present amounts to around 10 million ha, which is the largest in the history of Japan. In addition, the total volume of forest resources is at a record high level.

The process of afforestation has not been simple, and has given rise to many problems. The aim of this chapter is to explain this process and to outline problems in the development of forest-resource use in Japan over the last century. Further discussion is contained in Fujita (1989, 1991).

## Reorganisation of the ownership of the forest land in the Meiji era

Before the Meiji era (i.e. before 1868), most of mountain areas were mainly used for agriculture and as sources of fuelwood, timber and other forest products. Agricultural land was used for ordinary food crops, and areas of shifting cultivation were located around each settlement. Both permanent and shifting cultivation were geared mainly to the production of crops for subsistence, but a few crops were also produced for other purposes. Examples included tobacco and mitsumata (material for Japanese paper). Permanently cultivated fields were found near each village. In the mountain zones near plains, shifting cultivation was mainly located in the higher areas and in remote sites far from each mountain village. Extensive areas were used for shifting cultivation (Figure 12.2), which was carried out mainly by individual farmers.

Farther away from the villages, the mountain areas were used as a source of forest products and as grazing land for livestock, which produced fertilizer for the cultivated land. These more remote areas were, in general, managed collectively by villagers as common lands. Most of the mountainous zone, which was largely clothed in broadleaved forest, was used and managed in this way during this period. Small zones of more intensive commercial forestry were restricted to mountain areas near the Edo and Osaka markets. Industrial growth in the early twentieth century led to increasing demand for fuel, especially around cities such

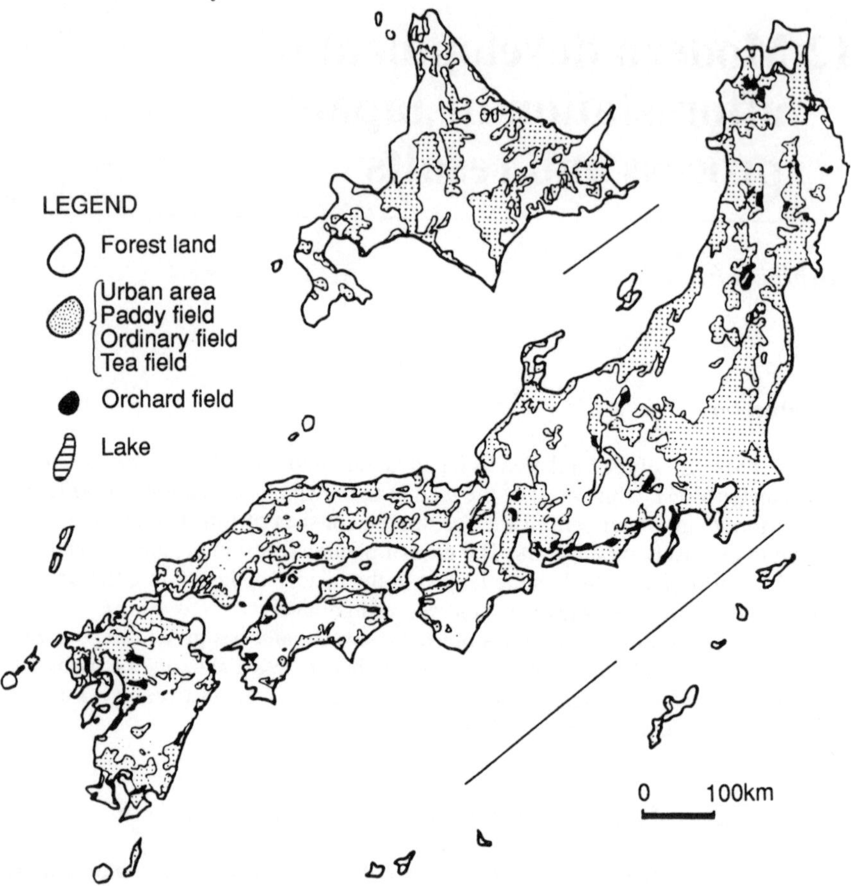


**Figure 12.1** Forest land and other land use in Japan.

as Tokyo, Osaka and Nagoya. As a result, the production of charcoal and fuelwood increased in some areas, and in some instances land previously used for shifting cultivation was given over to this use.

The pattern of use of cultivated fields, shifting-cultivation areas and intensive forestry zones continued until the early 1950s. The ownership of the forest land was not clearly established, and management was usually by clans.

The Meiji government enforced new land tax reforms, beginning in 1873, to support its financial base. At first, private ownership extended only to arable land, and the new owners had to pay tax according to the land prices that were created by the Meiji government. Later this reform extended to forest land. However, it was very difficult to establish the ownership of mountainous land, because a great deal of it was used in common by all the villagers. Eventually the Meiji government seized the communally owned forest land, most of which was under needle-leaf forest.

As a result the area of national (i.e. government-owned) forest land increased rapidly (Figure 12.3). Half of the forest area is now national forest. National

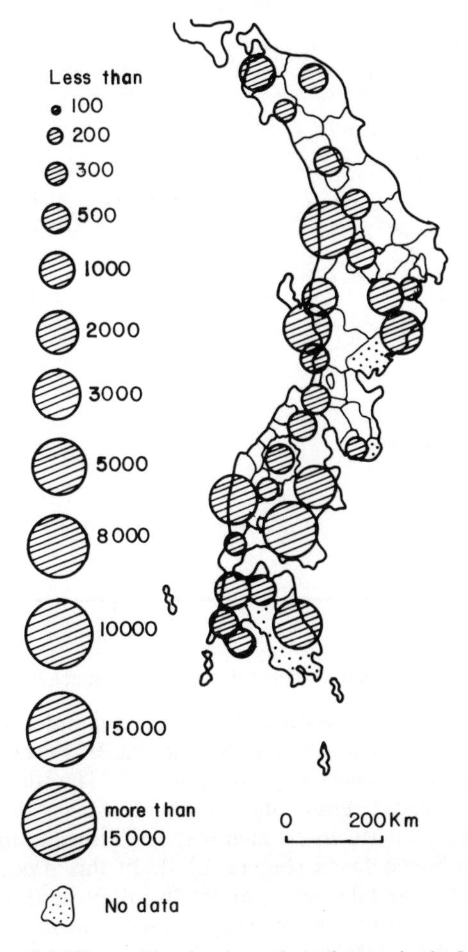


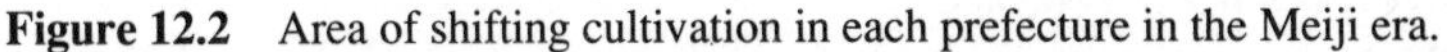

**Figure 12.2** Area of shifting cultivation in each prefecture in the Meiji era.

forests are located all over Japan, but are found especially in Hokkaido, Tohoku (northern area), Chubu (central area), southern Shikoku and southern Kyushu (southern area).

With this nationalisation, many farmers in these districts lost the forest land previously used for grazing and fuelwood. These farmers faced difficulties in supporting themselves following the Meiji government's new policy of restricting these traditional uses in the mountains, although at the same time many were employed as forestry workers. Conflicts emerged between government and farmers, especially when natural disasters occurred in the northern mountain areas. As a result, the government decided to restore the previous pattern of use and management on 10 per cent of the forest land, which satisfied some of the villagers.

A second policy concerning publicly owned forest land was then introduced. The new policy created publicly owned forest lands by seizing the rest of the common forest land traditionally used by the villagers. This was implemented

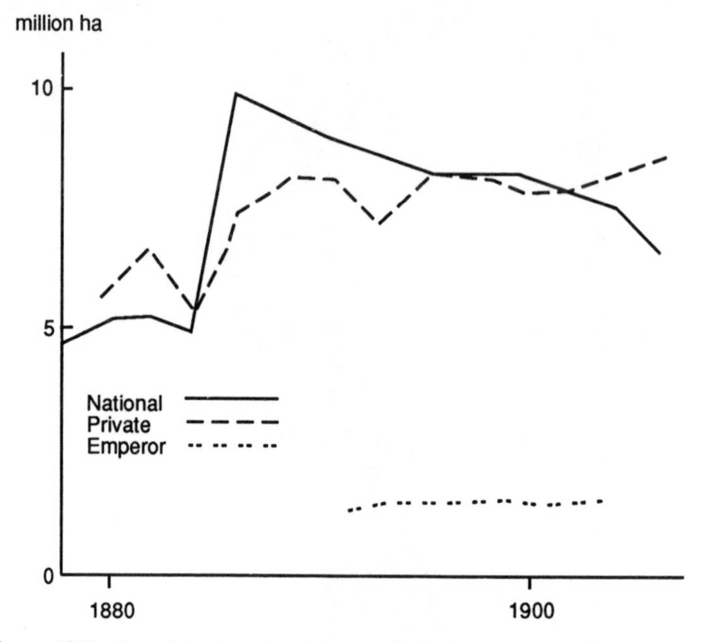


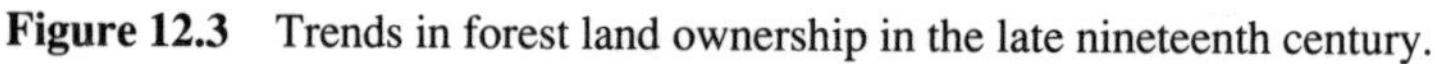

**Figure 12.3** Trends in forest land ownership in the late nineteenth century.

along with a policy that the government introduced in order to try to unify the small, traditional villages established in the Edo era. However, the newly established municipalities did not have any financial basis. Thus the government tried to create public forest lands to ensure the unification of the traditional villages. As a result, much publicly owned forest land was established at the expense of the traditional common forest lands (Figure 12.4). In this process, some of the traditional villagers opposed the new 'property ward' ownership system, because traditionally the group was the owner. They wanted to manage their forest lands by themselves using the old system.

During the Edo era the villagers' lives were closely connected with the forest lands in mountain areas. However, under the policies of the Meiji government, many villagers were forced to separate from the forest lands. As a result, many villagers were no longer self-supporting. They faced difficulties because of natural disasters and economic depression in the later years, which gave rise to many serious regional problems.

Thus, a variety of ownership systems such as national forest land, public forest land, 'property ward' forest land and private forest land were created under the Meiji government policy, and have survived to the present day.

## Forest management by the Meiji government

The Meiji government implemented its policy on national forest land by establishing a government forestry agency. From 1897, the government began to implement a full-scale national forest policy, based on the Prussian system of

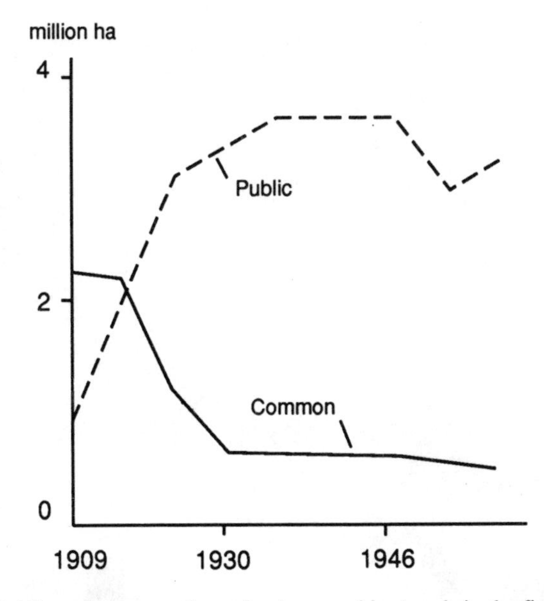


**Figure 12.4** Public and common forest land ownership: trends in the first half of the twentieth century.

management. More than 100 blocks of forest land were divided mainly on the basis of river basins, and the better forest lands were reserved as *Goryorin* (owned by emperor). The emperor was the largest owner in Japan before World War II.

The government refused to recognise the traditional rights of the villagers to enter forest land in order to cut wood and to gather grass. A new system of large-scale cutting and afforestation was introduced: this system was very different from the traditional one previously employed by the villagers. The whole concept of national forest land was alien to the traditional system of Japanese forest management.

### Development of the private forestry areas

On the other hand, traditional forest management systems had also been developed by the private owners of forest lands, after the Japanese–Russian War and World War I. Following these wars, timber prices rapidly rose in Japanese markets. Thus, larger owners of forest land, such as landlords, farmers and merchants, planted trees in previously unproductive woodlands or on areas of shifting cultivation. In such cases, they planted economic trees, sugi (Japanese cedar) and hinoki (Japanese cypress) using the traditional Japanese Yoshino forestry system.

This system had been developed along the Yoshino River, in response to market demand from Osaka, from the early years of the Edo era. The system is characterised by intensive planting using young plants raised from seed, and by

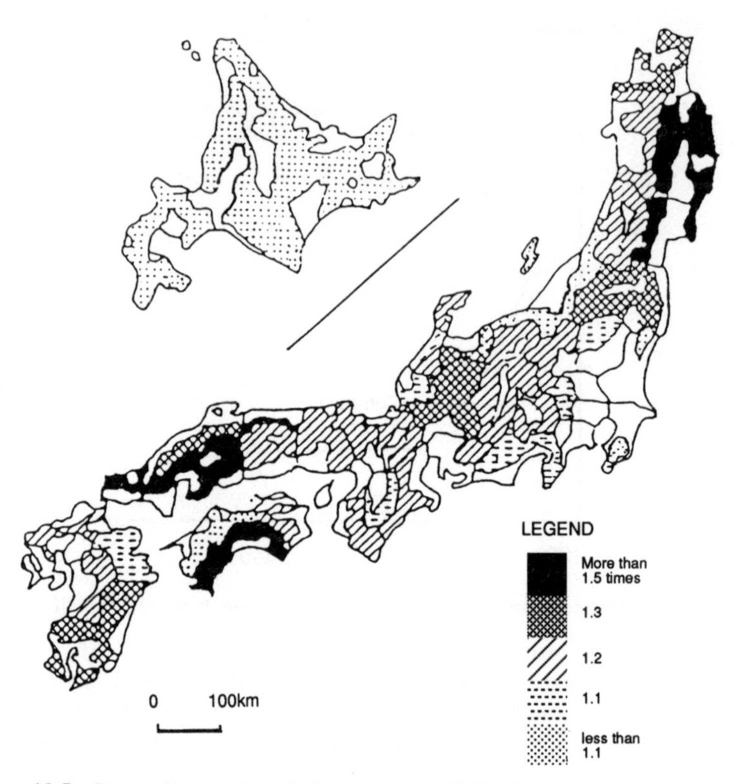


**Figure 12.5** Rates of expansion of planted forest, 1960–1970.

frequent cutting from 15 to 100 years. This system was good for the protection of the natural environment, especially against natural disasters, and was very different from the system of management imposed on the national forest lands. The Yoshino forestry system was also introduced into the management of the public-forest lands, especially by the municipalities and counties, but not by the prefectures.

Thus, there appeared two systems of forest management and afforestation – the government forestry system and the Yoshino forestry system. The two systems remained separate in later years.

The government had little interest in private forestry before World War II. However, it had recommended the setting up of an association of forestry managers and supported private afforestation, especially during the economic depression of the 1930s. In 1929, the government began to subsidise afforestation in private forestry. Both state and private forestry were scaled back by the financial problems of the government, especially during World War II. At the same time, private and military demand for wood and timber during World War II meant that Japan's forest resources were seriously depleted.

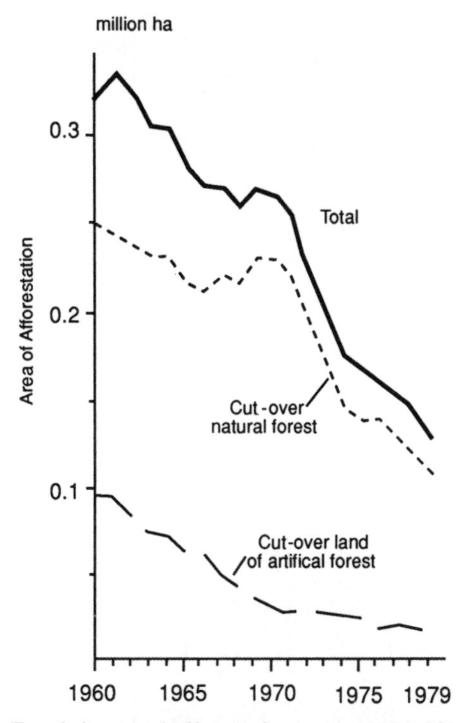


**Figure 12.6** Trends in annual afforestation rates, 1960–1980.

## Afforestation by mountain villagers

Just after World War II, the symptoms of deforestation were becoming all too apparent all over Japan. Large-scale disasters occurred in almost every river basin, as a result of extensive deforestation. Many cities had been damaged by air attacks during the war, and with increasing demand for timber, prices rose rapidly. Thus, an afforestation movement developed, aimed at rehabilitating forests all over Japan. This movement was supported not only by the large owners of forest lands, but also by medium-sized and small owners.

Two rebuilding movements were begun. One was inspired by the government, which encouraged tree planting by declaring special planting days and weeks. Tree-planting funds were established, and ceremonies were held each year, with the emperor attending each prefecture in turn. This campaign heightened popular awareness of the necessity to plant trees in order to prevent or control disasters.

The other movement was initiated directly by villagers in mountain areas. There was a widespread shortage of timber for rebuilding houses, factories and the other facilities damaged by bombing in World War II. In addition, during the Korean War timber was shipped from Japan to Korea as military material by the Americans and the United Nations forces. At the same time (around 1950), the initial post-war growth of the Japanese economy began, resulting in increasing demand for timber. The shortage of timber thus became severe, and timber prices increased rapidly. Many farmers began to plant trees, and government met half

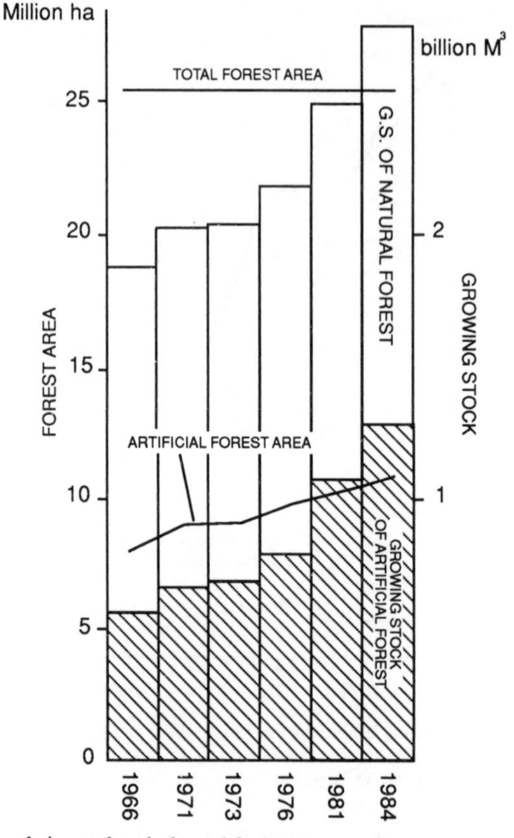


**Figure 12.7** Trends in total and planted forest areas and growing stocks.

the cost of afforestation. They mainly planted needle-leaf species such as sugi (Japanese cedar) and hinoki (Japanese cypress), because these were the best for constructional materials: broadleaved forest was in places thus replaced by needle-leaf forest. Planting extended into areas previously used as sources of charcoal and fuelwood.

Thus, much of the previously deforested area had been afforested by 1960. The volume of growing stock did not increase, because of the marked growth in demand for timber, but the occurrence of flood disasters decreased rapidly with the re-establishment of forests in the mountains. The pattern of forest expansion is illustrated in Figure 12.5.

Afforestation could not, in the short term, solve the immediate problem of timber shortage. In response to this shortage, the government decided in 1963 to import timber from the USA, the USSR and south Asian countries, and by 1970, imports accounted for more than half of Japanese wood consumption. As a result, timber prices fell by half, and the production and the annual afforestation rate gradually decreased (Figure 12.6).

Very different circumstances prevailed in the national forest area. National forest management was reorganised to include the large forest lands in Hokkaido.

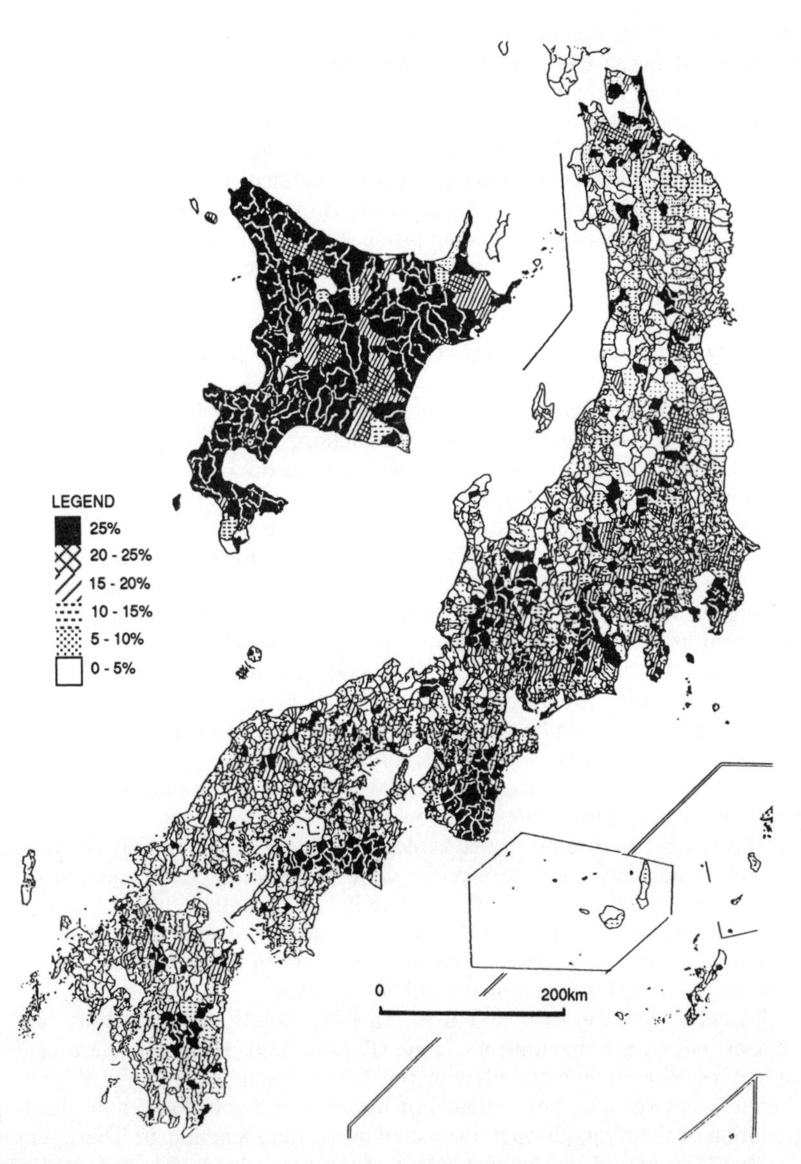


**Figure 12.8** Proportions of absentee land owners in each municipality in mountain areas, 1980.

A financially self-supporting system of accounting was introduced, and was intended to be quite separate from general government accounting. During the period of rising timber prices, the management of national forest lands was easy and good. However, with decreasing timber prices, the national forest-land agency had to increase cutting rates in order to support itself financially. At first, the replanting of cut-over land of natural and artificial forests meant that afforestation rates increased. However, the high rates of afforestation could not be sustained, because of rapidly increasing costs. Moreover, deforestation affected many mountain areas, and natural disasters occurred. As a result, calls for the preservation of natural forest resources in the national forest lands were voiced.

In the period after World War II, the climate for forestry and afforestation changed rapidly. Mountain villagers quickly adopted forestry-management techniques, especially on private forest land. However, when timber imports increased rapidly and prices for home-produced timber fell, many villagers gave up forest production and became estranged from the forest.

## Forestry depression and forest resources: current problems

More than 25 years have passed since the rapid increase in timber imports. As a result of decreased domestic forest production, many villagers left their rural homes. They moved to urban areas where industries and services were being developed and where there was a strong demand for labourers. As a result, mountain areas increasingly experienced depopulation.

While the volume of growing stock rose with the maturing of post-war plantations (Figure 12. 7), much of the forest resource was neglected and indeed was turning to waste land in the absence of care and management. Most of the neglected land is located in the more remote areas, where the forest was previously used for the production of fuelwood and which was then changed to 'artificial' forest. These areas could not be worked or managed economically, because of the fall in timber prices and the depression in the forest industry. Many people have left their homes and 'socially formed empty areas' (a term coined by the author) developed. Sharecropping systems were introduced, but failed because of the shortage of forestry labourers.

The number of absentee forest landowners has increased rapidly (Figure 12.8), and forest lands no longer receive the same degree of attention and management from the mountain villagers as they formerly did. Mountain villagers now work in other occupations. These remote mountain areas are extensive, and are very important for forest functions such as the support of water resources, protection from natural disaster, and tourism and health resorts.

Recently, non-forestry land uses, such as hotels, skiing resorts and golf courses, have been invading this zone (Figure 12.9). However, these land uses cannot be directly integrated with the forest resources. New problems have therefore emerged in these areas. For instance, the golf courses are leading to pollution of the rivers through the use of agricultural chemicals. These problems are due to a lack of the understanding of the forest lands. There is at the same time, however, a growing movement for the preservation of the natural environment on the part of urban residents.

In Japan, the establishment of the vast national forest lands excluded the mountain villagers from some mountain areas before World War II, and the large volume of timber imports, and falling prices for home-produced timber, had a similar effect on some of the private forest areas after the war. As a result, both types of areas are deteriorating through the lack of effective management, even though they contain some of the country's best forest resources.

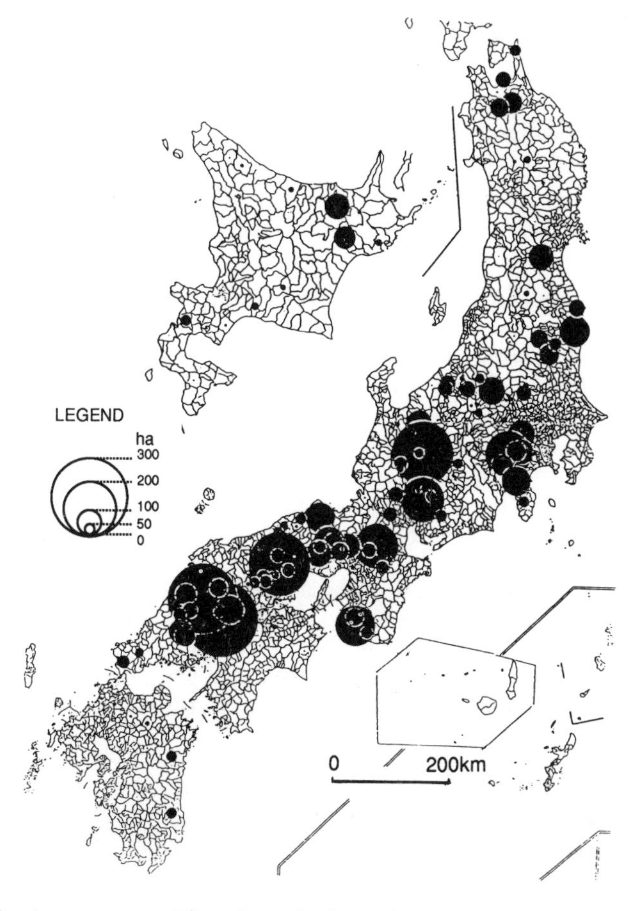


**Figure 12.9** Areas converted from forest lands to other uses, by municipalities, 1980.

Thus, how to manage and how to preserve these run-down forest lands are serious problems. The problems are the result of the Japanese policies, and of a change in perception of the forest environment over the last hundred years. Extensive afforestation has been achieved, but difficulties have arisen over the management and use of the planted areas. We need to develop a new attitude towards our forest resources, and seek new approaches to using forest lands.

## References

Fujita, Y. (1989) 'Land-use change in the mountains of Japan since the period of higher economic growth', in R.D. Hill (ed.) *Land-use change*. Hong Kong University Press, Hong Kong, pp 85–100.

Fujita, Y. (1991) 'The change of forest land use since the Meiji Era in Japan', in I. Ota (ed.) *Environmental change and GIS*, Hokkaido University, Asahikawa, pp 347–54.

# 13 Afforestation: the United States

*Michael Williams*

Among the multiplicity of factors at work in the maintenance of a forest cover on any part of the globe, afforestation seems, on the face of it, to be simple, straightforward, purposefully motivated and easily isolated for treatment. But a moment's reflection reveals how difficult it is to separate the topic of replanting from wider and bigger forest issues that may be none of these things. For example, while afforestation is often a response to constructional timber shortages, often it is also a response to changing climate, ideas of climatic amelioration, or fuelwood deficiency. In addition, afforestation, which aims to increase timber growth, cannot be separated from the wider issues of forest management, and therefore, it cannot be divorced entirely from fire control, stand treatment, eradication of disease, and raw material demand and substitution, all of which aim to protect and enhance existing timber stands. Finally, the interplay of different national policies and economic priorities at different times create conditions that are either favourable or unfavourable not only to the planning and progress of afforestation but also to natural forest growth. Agricultural expansion usually diminishes the forest area, but agricultural contraction allows it to expand.

All these themes are represented in the story of afforestation in the United States, a continental and dynamic economy and society which, during the last two hundred years, has moved from being parochial and small-scale to global and large-scale. Yet, because of the abundance of timber and ecological vitality of the stands the idea of purposeful afforestation has not loomed large. With an area in the mid-twentieth century of (in round figures) 500 million acres (202.3 million ha) of commercial forest land, and 250 million acres (101.2 Mha) of non-commercial forest land, to say nothing of the 114 million acres (46.1 Mha) of forest probably cleared for agriculture before 1850 and the further 190 million acres (76.9 Mha) cleared from mid-century to 1910, the necessity to plant more trees has not seemed pressing. Nevertheless, trees were planted in large quantities, initially for their supposed climatic and general environmental effects, and later in order to replenish spent stocks in prime timber-producing areas.

The emphases and shifts of policy differed from time to time and therefore are best looked at chronologically.

## The nineteenth century

The juggernaut progress of agricultural settlement and associated clearing in the eastern United States went on unabated during the nineteenth century, reaching a peak during the decade 1870–79 when nearly 50 milion acres (20.3 Mha) were converted to agricultural land (Williams 1989). Towards the end of the century the fear grew that clearing might be exceeding growth and that soon a timber shortage would ensue. Nevertheless, this had little tangible result in terms of afforestation because of the abundance of natural forests, although it was an essential backdrop to, and motive for, debate and legislation about preservation, management and protection of the nation's forests. By 1890 the Forest Reserve Act had been passed which gave the president power to set aside land 'in any State or territory having public land bearing forests' (US Congress, House 1890). By 1910, 200 million acres (80.9 Mha) had been declared National Forests, wholly in the western states (US Bureau of the Census (USBC) 1977).

Thus, it seems very likely that the abundance of timber and the knowledge that such a vast area of forest had been set aside for public purposes took the urgency out of afforestation for timber supply. In practical as well as in perceptional terms, the need to plant trees did not seem to apply.

### *Afforestation to change the climate*

However, in the Midwest interest in planting trees loomed large. By 1830 agricultural settlement had reached scattered open prairie patches of north-west Indiana, south Michigan and south-west Wisconsin, which merged into true prairie further west in Illinois. Generally the treeless prairies proper were not encountered until after the Civil War. Then expansion posed questions of timber and fuel supplies in a way that had not occurred before. Resources had to be re-evaluated. Initially there was a prejudice against the non-treed, open areas because it was common view that if tree growth indicated fertility (and most people thought that it did) then the prairies were useless. The lack of timber for fuel, fences and construction was made up for either by 'imports' from the timber-rich lake states to the north (and later from the South), or by securing ownership of distant timber lots on river frontages. Where these were not available settlers began to plant trees and create woodlots (Birch 1974; Jordan 1964; McManis 1964).

The myth of the 'Garden of the World' had come up against the equally vibrant myth of the 'Great American Desert'. How could the 'Desert' be transformed into the 'Garden'? By perverse reasoning, the idea that forest destruction caused a decrease of rainfall was turned on its head to imply that forest planting would result in greater rainfall (Smith 1947; 1950; Emmons 1971a; 1971b; Kollmorgen and Kollmorgen 1973). The idea was much older than the settlement of the Plains and can be traced back, at least, to the writing of Buffon in 1789, but it was picked up with vigour by many prominent American scientists and populists, like Frederick V. Hayden of the Federal Geological Survey, Joseph Henry of the Smithsonian Institution, and others (Emmons 1971b; Hayden 1867). When this advocacy was combined with the official pronouncement by Joseph Wilson, the influential Commissioner of the General Land Office, that every settler be required to plant trees in order to change the climate, the result was electric (Wilson 1868). These 'boosterish' attitudes were aided by the creation of Arbor Day in 1872. Tree-planting ceremonies by school children and citizen's groups took on a quasi-religious tone and it seemed as though the transformation of the plains was at hand (Olsen 1972a; 1972b).

The outcome was the 1873 Timber Culture Act which offered 160 acres (65 ha) to homesteaders who planted a designated percentage of their land in trees (US Congress 1872). The Act was in operation for 18 years during which time some 40 million acres (16.2 Mha) were entered as applied for, of which 9.8 million acres (4.0 Mha) were patented, mainly in a block of adjacent Plains states; 2.5 million acres (1.0 Mha) in Nebraska, 2 million acres (0.8 Mha) in Kansas, 2.1 million acres (0.8 Mha) in South Dakota, 1.2 million acres (0.5 Mha) in North Dakota and 0.6 million acres (0.2 Mha) in Colorado. However, it is doubtful if even a quarter of a million acres (101,000 ha) were planted with trees (Hibbard 1965; Kollmorgen 1969; Raney 1919; McIntosh 1975).

The truth was that the Act had been used extensively as a means of speculation; eastern speculators bought to sell to intending settlers, cattlemen to secure their water frontages, and hem in other land holders, and everyone bought to enlarge their holdings. By 1882 the Commissioner of the General Land Office called attention to the abuses of the Act and said: 'my information is that no trees are to be seen over vast regions of the country where the timber culture entries have been most numerous' (US Dept. of the Interior, 1884). By 1891 the abuse was so obvious that the Act was repealed.

Thus, America's first great experiment in afforestation ended largely in failure, although many farmsteads planted shelter belts around their houses and woodlots adjacent to the farm. Despite repeated attempts by Nebraskan boosters like Samuel Aughey, Charles Wilber and Charles Bessey to revive the idea during the 1890s, the idea died for lack of a scientific basis (Aughey and Wilbur 1880; Kollmorgen 1969; Kollmorgen and Kollmorgen 1973; Overfield 1979). The exceptionally dry years of 1889–90 showed that planting in this part of the country did nothing to ameliorate the climate.

## The twentieth century

The years since the beginning of this century have seen an unprecedented increase in the stock of timber growing in the United States compared to earlier years. This has been the result of two processes. Firstly, there has been spontaneous reforestation with the natural regeneration of forest on farmland that has been abandoned, and land that has already been logged. To a very large extent Clawson's assertion that 'timber growth is a function of timber harvest' is true (Clawson 1979). Secondly, there has been purposeful afforestation with the planting of trees to restock the depleted forest areas in regions of commercial lumbering and marginal farming.

### *Reforestation through reversion*

In the past, clearing the forest for farming has always left a large amount of land in trees. The percentage of land in forest rarely falls below half over much of the southern states or below a quarter to a third in the more densely settled Middle Atlantic and New England states. Thus, there was a stock of vigorous trees ready to reseed and recolonize the farmland once it had been abandoned.

Abandonment and regrowth was commented on as early as 1840 in New England, and by 1880 the process had spread southwards through the Middle Atlantic states of New York, Pennsylvania and New Jersey. By 1910 it had penetrated the east-central states of Ohio, Indiana and West Virginia. In later years, and especially after 1930, abandonment has been common throughout the South as old cotton and tobacco fields revert to pine forest (Hendrickson 1933; Reuss *et al.* 1949; USBC 1920; Vaughan 1928; Wilson 1936).

There has been a multiplicity of reasons for this shift in the land use and land cover. In New England and the Middle Atlantic states the land was often steep, stony, infertile and eroded easily, and it had only been utilized with the application of immense human effort. Much of it was also climatically marginal for many staple crops. They were poor lands when compared to the new lands opening up in the Midwest and West where terrain, soils and climate were much more favourable to mechanized agricultural production.

In the South the situation was somewhat different. The long-hallowed practice of, in effect, shifting cultivation in the forest in order to combat the erosion and nutrient loss from growing the demanding crops of cotton and tobacco was lengthened with the mass out-migration of black labour to the northern industrial cities during the 1930s to 1950s. Additionally, the shift of agriculture from the Piedmont of South Carolina, Georgia and Alabama to the coastal plain and the new irrigated areas further west added to the outright abandonment. For example, whereas there were 4.1 million acres (1.6 Mha) of cotton in 88 counties in the Piedmont in 1920, there were only 0.3 million acres (121,000 ha) left by 1959. In Piedmont Virginia, tobacco fell from 172,000 acres (69,000 ha) to 73,000 acres (29,000 ha) in 29 major tobacco counties, and overall there was the abandonment of 1.3 million acres (526,000 ha) previously cleared from the forests (Hart 1968; Vance 1932).

Regrowth of the forest is not easily detected as it goes unrecorded in the agricultural census and can be deduced only by comparing 'total farmland' with 'farmland (less farm woodland)', and then only with any reasonable precision in the predominantly wooded counties of the 31 eastern states. Even then there are complications. For example, the land cleared in one period could have reverted to forest and then been cleared again at a later date, or the land abandoned to forest could be compensated for by clearing elsewhere in the forest. Again, aggregate statistics can be misleading because woodland already covers at least a quarter of the total area. Elsewhere, some abandoned farm land will be taken over for urban and industrial expansion, freeway construction, strip-mining, and therefore never revert to forest.

Generally, in a country imbued with concepts of progress and development, reversion denoted backwardness and did not figure prominently in public and official pronouncements. A series of major inquiries and reports from 1920 to 1940 into timber supplies and the lumber industry were hallmarks or stages in thinking about the forest, as well as declarations of intention over future policy. The first in 1920, the Capper Report (complied mainly by William Greeley), barely recognized the phenomenon of reversion, as it emphasized the 'five *D*s' – devastation, depletion, deterioration, decay and disappearance – although the growth of the New England forest by over 13 per cent was acknowledged (US Forest Service (USFS) 1920). A little later, Greeley saw that a sort of equilibrium was occurring in the American forest, and by 1925 could say:

> While the inroads of the farm are continuing here and there, the great tide of forest clearing for cultivation seems largely to have spent itself. For many years, indeed, the abandonment of farm land in forest growing regions of the old States has practically offset new clearing in the agricultural frontier (Greeley 1925).

Recognition of the shift went one step further in the Copeland Report of 1933, with a detailed discussion of abandonment by C.I. Hendrickson. He calculated that 2.3 million acres (0.9 Mha) had been lost in New England between 1880 and 1930 alone and a further 0.77 million acres (311,600 ha) in the Middle Atlantic states. He calculated the net abandonment totalled over 19.1 million acres (7.7 Mha) between 1910 and 1930 – in fact it was nearer 13 million acres (5.2 Mha) – and he foresaw even greater changes with rural to urban migration, migration to the West, and the increase of mechanization, which would precipitate the decline of horses and mules and their replacement by tractors and the consequent release of up to 15 million acres (6.0 Mha) of grass and forage crops. In all, he expected a further 21.5 million acres ( 8.7 Mha) currently designated as idle or fallow in the 'first stages of abandonment' to be included in the total as well as a further 29 million acres (11.7 Mha) of 'pasture other than plowable or woodland pasture', much of which was in an advanced stage of abandonment. Therefore, if only a half of these estimates were added to the 19.1 million acres (7.7 Mha), there would be nearly 40 million acres (16.2 Mha) immediately available for afforestation and possibly between 75 and 80 million acres (30.3–32.4 Mha) by 1950. These optimistic calculations were instrumental in influencing and encouraging the compilers of the 1933 Copeland Report to recommend vigorous and extensive state intervention in the private forest industry and a vast programme of replanting (Hendrickson 1933).

The reality was not quite what had been envisaged by the enthusiastic interventionists of the 1930s, buoyed up as they were by the rhetoric of the New Deal in order to mitigate the harsh reality of the Depression. The years of the decade saw a slight gain in the amount of forest cleared, although the picture varied enormously throughout the country. In the South, detailed surveys covering 283 million acres (114.5 Mha) of forest revealed that abandonment was running at somewhat more than half the rate of clearing (2.0 to 5.3 million acres – 0.8 to 2.1 Mha) and that it dominated all areas except North Carolina, North Alabama, and Mississippi. During the first half of the next decade, war-time emergency caused a marked up-swing in clearing, so that more farmland was created out of the forest than at any other time in the nation's history, and there was the destruction of much high-yielding forest. Since then, however, the trend has been downward, with the net losses of cleared land due to all causes totalling 79 million acres (32 Mha) since 1940 (Reuss *et al.* 1949).

Against only partially accurately calculated and surveyed estimates of abandonment and therefore the intuitive judgements and guesses of the reports, one can set the realities of the census data for the 31 states of the eastern portion of the country, where losses and gains can be calculated reasonably accurately (Table 13.1)

Net loss of cleared farmland between 1910 and 1959 was 43.8 million acres (17.7 Mha) with 65.5 million acres (26.5 Mha) having been abandoned and 21.7 million acres (8.8 Mha) gained, mainly from the forests of Florida and coastal

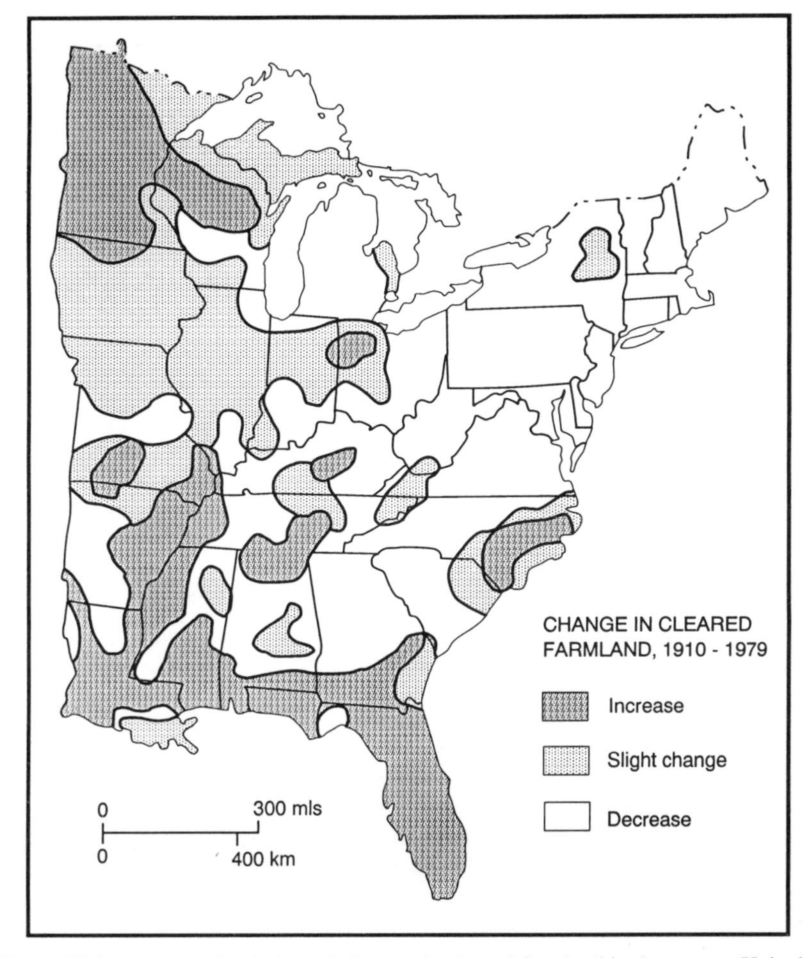


**Figure 13.1** A generalized view of changes in cleared farmland in the eastern United States, 1910–1979. (Based on Hart (1968) and US census data.)

Alabama (Figure 13.1). From 1959 to 1979 another 16.9 million acres (6.8 Mha) were lost to agriculture or gained from the forest, much of this in the Mississippi bottomlands (Abernethy and Turner 1987; MacDonald *et al.* 1978). In other words, 865,000 acres (350,000 ha) of once-cleared farmland were lost to agriculture and added to the forest every year between 1910 and 1959 and a slightly lesser amount of 845,000 acres (342,000 ha) has been added to 1975. The indications are that the rate has slowed down considerably since then (Clawson 1981; Hart 1968; Reuss *et al.* 1949; USFS 1990).

How much of this abandonment actually reverted to forest will never be known, but spot checks in the South using air photographs during the 37 years between 1937 and 1974 show that the shift from agricultural use to forest use exceeds that from agricultural use to all other land uses by a ratio of 15:1. In the

**Table 13.1** Cleared farm land in the United States, 1910–1979 (millions of acres)*

| | Coterminous US | Thirty-one eastern states | | |
|---|---|---|---|---|
| | Total farm land | Total farm land | Farm woodland | Cleared farm land |
| 1979 | ? | 351.1 | 72.5 | 278.6 |
| 1975 | 1013.7 | 343.4 | 77.7 | 265.7 |
| 1969 | 1059.6 | 369.4 | 78.2 | 291.2 |
| 1965 | 1106.9 | 396.1 | 95.6 | 300.5 |
| 1959 | 1120.2 | 415.4 | 113.2 | 302.2 |
| 1954 | 1158.2 | 455.0 | 133.5 | 321.5 |
| 1950 | 1161.4 | 470.5 | 135.9 | 334.6 |
| 1945 | 1141.6 | 467.6 | 118.5 | 349.1 |
| 1940 | 1065.1 | 459.9 | 106.1 | 353.8 |
| 1935 | 1054.5 | 474.5 | 132.2 | 343.3 |
| 1930 | 990.1 | 443.4 | 110.1 | 333.3 |
| 1925 | 924.3 | 445.7 | 108.7 | 337.0 |
| 1920 | 958.7 | 478.0 | 130.4 | 347.6 |
| 1910 | 881.4 | 490.3 | 144.3 | 346.0 |

*1 million acres = 404,680 ha

*Source*: Hart (1968) and an analysis of the *US Agricultural Census* for 1965, 1969, 1975 and 1979.

countervailing trend from forest to all other land uses (built-up and agriculture), as against from agriculture to all other land uses, the ratio was remarkably similar at 16.5: 1. Change, therefore, has been overwhelmingly towards the abandoned land reverting to forest, and urban encroachment, except around the large eastern seaboard metropolises, has not been great (Hart 1980).

The change has been scattered, incremental and difficult to detect. It is, however, revealed in a number of ways – from gross state and national statistics, from small-scale spot surveys and particularly from a glance from the window of an aircraft. The mosaic of different shades of green in the forest canopy, particularly in the even-growing pines of the South, has remarkably rectangular boundaries. They usually represent forests in various stages of spontaneous regrowth and not purposeful afforestation as first seems the case. In 1848, George Perkins Marsh, the celebrated author of *Man and Nature*, warned the farmers of Rutland County in Vermont at their annual fair that if any one of them in middle age returned to their birth place the ravages of deforestation, eroded slopes, and dried-up water courses would mean that he looked upon ‘another landscape than that which formed the theatre of his youthful toil and pleasures’ (Marsh 1848). Now, the reverse has happened. Any contemporary American returning to their birthplace on the east coast after 30 or 40 years’ absence would also find ‘another landscape’, but this landscape would be unrecognizable because of its heavy greenery, limited views and lack of farming activity.

### *Afforestation through tree planting*

It is often forgotten that unless a forest has been completely stripped of all its trees, it will (especially in temperate latitudes) regenerate naturally and fairly

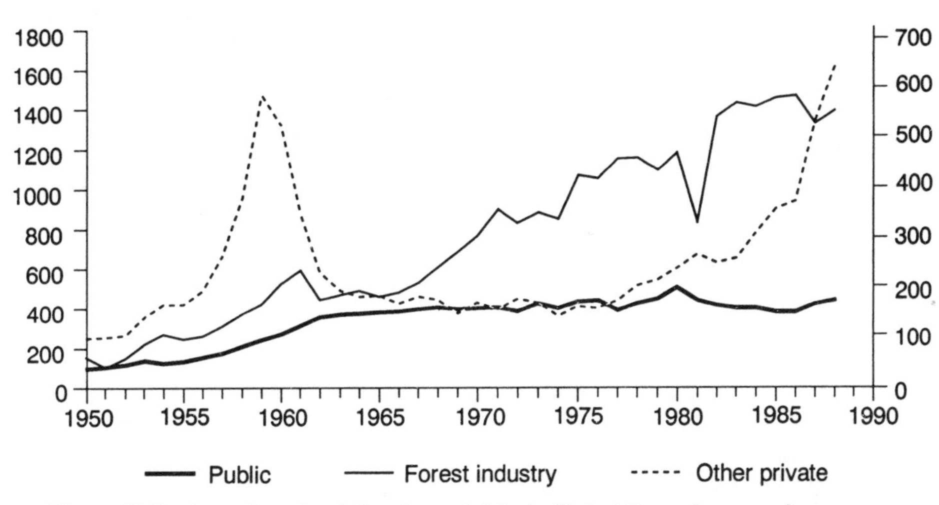


**Figure 13.2** Area planted and directly seeded in the United States, by type of ownership, 1950–1988. (Based on USFS 1990.)

rapidly. However, natural regeneration may not result in species of the desired type. Over large areas of the South, for example, the natural succession is to less productive hardwoods, unless positive steps are taken to encourage pine regeneration.

In addition, the natural process of regeneration may be too slow and uncertain for modern forestry practice, especially where there is an investment of fixed capital in the form of mills, settlements and transport infrastructure. Consequently the area planted to trees and artificially seeded has increased at an accelerated rate in order to get a better control of species, spacing and stocking levels, and to allow for the introduction of genetically improved stock.

The massive national afforestation programme of 50 million acres (20.3 Mha) suggested by the Copeland Report of 1933 will probably never be fulfilled, if for no other reason than that it had not taken reversion into account, but replanting and reseeding have been greater than acknowledged. Under the Clarke–McNary Act of 1924, small appropriations were made from federal budgets for the growing, distribution and planting of timber stock for farmers in order for them to extend shelterbelts and manage their farm woodlots more efficiently. The next major federal intervention, the Knutson–Vandenburg Act of 1930, provided funds for planting trees in National Forests and cut-over areas in the hands of private foresters. With these acts and various subsequent farm programmes under the New Deal, particularly the activities of the Civilian Conservation Corps (CCC) (Salmond 1967), the groundwork was laid for a degree of federal–state cooperation in the distribution of trees and seedlings for forestry and wind-break purposes. The Cooperative Forest Management Act of 1950 was the culmination of this process (Dana and Fairfax 1980). In all, during the 31 years between 1939 and 1970 at least 12.6 billion trees were distributed and 14.8 million acres (6.0 Mha) planted, and probably about 1.7 million owners affected.

Impressive as this was, it was only a portion of what was done. A variety of other agricultural programmes, such as the Agricultural Conservation Program (1930), the Conservation Reserve Soil Bank Program (1956–61) and its successor the Conservation Soil Bank Program (1960), together with the Forestry Incentives Program, added enormously to the area planted, which up to 1955 never fell below 1 million acres (0.4 Mha) (Figure 13.2). Up until the 1950s the main priorities were for shelterbelts on the Great Plains (which was one of the 'grand' projects of the New Deal), forests on badly eroded land, on the unstable sandhill country in Nebraska, Kansas, and South Dakota, and in silt-choked catchments everywhere (Droze 1977; USBC 1977; Wessel 1967). During the 1950s the emphasis shifted overwhelmingly to farms and other private ownerships, particularly in the South (Clark 1984), where payments were made under the Soil-Bank Program to farmers to retire land from crop use. Over 6 million acres (2.4 Mha) of farmland were affected between 1956 and 1961.

When the Soil Bank Program came to an end planting continued but at a much reduced rate of about 0.5 million acres (202,000 ha) per annum until the mid-1980s, when it rose again to 1.6 million acres (0.6 Mha) a year under the stimulus of a great variety of new programmes to encourage tree planting, targeted at private ownerships, and taken up overwhelmingly in the marginal cropland in the South.

Examples of this legislation include the Renewable Resources Extension Act (RREA), which provided funds for technical and educational assistance for landowners to protect and manage their land, over two-thirds of the funds going on forestry. The Cooperative Forestry Assistance Act (1978) specified state and federal matching funds for nursery production and tree improvement schemes. This was further extended by the Federal Incentives Program (FIP) and the Agricultural Conservation Program (ACP) which, although concerned primarily with water and soil conservation, have provided 65–75 per cent of the money for tree planting, site preparation, timber stand improvement, and natural regeneration. These have proved very popular and have accounted for 40–50 per cent of all afforestation activity. Some of the marked upsurge of afforestation since 1985 can be attributed to activity under the Conservation Reserve Program established under the Food Security Act of 1985. Under this Act, farmers can put land highly susceptible to erosion into reserve and receive ten annual rental payments for it, and up to 50 per cent of the costs of planting trees. This has proved very attractive as some trees are ready for an initial harvest after only five years. In addition to all these federal and state programmes, tax policies since 1980 in the shape of incentives and credits, and amortization provisions have recognized the long-term nature of planting investments, and attracted outside investment capital (USFS 1990; Risbrut and Elleson 1983; Royer and Moulton 1987).

While farmer and other private plantings have suddenly sprung back into prominence, the most steady increase in planting during the last 30 years has come from the lumber industry and not as a part of any federally instituted farm programme. The vast majority of lumber companies are painfully aware that the past practice of ruthless cutting, the abandonment of old plant and settlements, and moving on to new stands (even if there were any) is an expensive business. About 1.0–1.4 million acres (0.4–0.6 Mha) a year have been planted. It is cheaper to replant forests, improve fire control, and engage in general forest management.

It is known that the public and private sectors spend over £2 billion a year on management, research and assistance. How that is broken down for the industry as a whole is not known but we do know that half of this is USFS expenditure, of which approximately 75 per cent goes on general management and road construction (roughly equally), 8.4 per cent on fire protection, 8.2 per cent on research, 5.7 per cent on assistance to states and private owners, and 2.6 per cent on insect control. Other than the assistance element, one must assume that roughly the same breakdown occurs for the private lumber industry (USFS 1982).

Lumber industry nurseries have multiplied enormously in the post-war years, as have commercial and federal- or state-owned nurseries. Nearly are all are stocked with genetically improved varieties of trees. The emphasis in afforestation has swung back and forth from planting with artificially raised trees, to direct seeding and back again to planting. Direct seeding was carried out with some success in the southern pine forests where two-thirds of the 1.2 million acres (0.5 Mha) were seeded, and the Pacific coast where about a quarter of the 0.5 million acres (0.2 Mha) were seeded. However, success in seeding depends on adequate site preparation, especially by using pesticides and herbicides to keep down rodents and insects, and competing vegetation. Increasingly, environmental constraints on the use of chemicals are causing problems and the emphasis has now swung back to planting, which has a greater rate of survival (USFS 1982).

It is often assumed that once an area is reforested, all of it grows to maturity, but the truth is that there is always a mortality of trees, which at times can be devastating. Exact survival rates are difficult to come by but various estimates suggest that it is 70–75 per cent in the fast-growing warm South, and lower in other areas of the continent. One study of the Yazoo–Little Tallahatchie Flood Prevention Program in Mississippi, where nearly 800,000 acres (300,000 ha) were planted in blocks owned by small-scale holders, has shown that it was necessary to replant nearly one-fifth of the land during the subsequent 30 years. Success depends on rodent and disease eradication (Williston 1979).

When all is said and done, however, the area affected annually by planting and seeding is only a very small proportion of the commercial forest – a mere 0.78 per cent even at the current high rate of afforestation. However, there are significant differences in afforestation according to ownership, approximately 1.6 per cent of commercial timberland being affected, 0.3 per cent of farmer timberland and only 0.4 per cent of National Forests.

### *Other management techniques*

As mentioned in the introduction to this chapter, afforestation is only one strand in achieving a greater yield from the forest resources of a country. The steady increase in the yield of softwoods from 15.2 cubic feet per acre in 1952 to 26.6 cubic feet per acre in 1987, and for hardwoods from 12.1 to 20 cubic feet per acre during the same period, could not have come about without various management techniques, such as fire control, intermediate stand treatment, and disease control, particularly in the regenerated regrowth forests on the old farmlands (USFS 1990).

For example, the intermediate stand treatments whereby forest species composition is changed between regeneration and harvest, defective trees are culled which increases the growth of more favoured trees, and seedlings of desired species are released in recently regenerated areas, all boost production. This has occurred in some 1.5 million acres (0.6 Mha) a year. A similar effect is achieved through fertilizing, draining, and the thinning of low-value timber and pruning of lower branches. Fertilizing has been particularly effective in the nitrogen-deficient soils of the Douglas-fir growing region of the Pacific Northwest and the poorly drained phosphorus- and nitrogen-deficient soils of the pine-growing areas of the Southern Coastal Plains. Yields have risen by 200 to 800 gross cubic feet per acre over a ten-year period in the Northwest (USFS 1990).

In a similar fashion the reduction of losses through the elimination of fire, insects and disease can boost production. For example, the rapid salvage of dead or dying trees (especially after windthrow or fire) has contributed to annual growth and timber inventories. Fire prevention and control are the oldest, most widespread and probably the most efficient managerial effort in US forestry. It would be true to say that nearly all the 483 million acres (195.4 Mha) of commercial forests are protected and that the area burnt annually has fallen from 30–40 million acres (12–16 Mha) at the beginning of the century to 3–5 million acres (1.2–2.0 Mha) in the mid-1980s. However, controlled burning to reduce combustible material has been hampered in recent decades by air-quality constraints, and the greater intermingling of urban structures within forested areas (USFS 1990; Williams 1989).

Finally, there is a vast expenditure on disease and insect control and eradication. Southern pine beetle, mountain pine beetle, and western spruce budworm are just a few of the many organisms that inhabit the natural forest ecosystem and have to be kept at bay. Fustiform in the southern pines is killing or deforming millions of slash and loblolly pines and only the planting of genetically resistant stock as replacements or in afforestation schemes seems to be able to resist it.

### *Afforestation: the twenty-first century*

Forests are slow growing and reasonably predictable in their behaviour, so that provided certain assumptions like population change, personal income, inflation, agricultural productivity, etc., are observed, it is possible to project trends forward for half a century with some degree of confidence. Thus, in 1989, the USFS completed its now statutory ten-year Renewable Resources Assessment (under the Forest and Rangeland Renewable Resources Planning Act of 1974) and took a look at possible futures to 2040 (USFS 1990).

It is predicted that the area of commercial forest will continue the decline already noted between 1962 and 1987, when it slipped from a peak of 515.1 million acres (208.5 Mha) to 483.2 million acres (195.5 Mha). It may fall a further 20 million acres (8 Mha) to 462.8 million acres (187.3 Mha) by 2040 (Table 13.2) The decline is accounted for by clearing for transport, housing and industry but also to the large area of public land withdrawn from commercial forests for parks, wilderness, and recreational uses, particularly in the West (Waddell *et al.* 1989). Thus, the forest is not physically eliminated but lost to commercial use. This trend would be greater but for the continued addition of

**Table 13.2** Area of timber land in the United States, by ownership and region, 1952–1987, with projections to 2040 (millions of acres)

| | | | | | | Projections | | | | |
|---|---|---|---|---|---|---|---|---|---|---|
| | 1952 | 1962 | 1970 | 1977 | 1987 | 2000 | 2010 | 2020 | 2030 | 2040 |
| Ownership | | | | | | | | | | |
| Public | 152.8 | 152.5 | 150.2 | 144.2 | 136.3 | 134.3 | 134.3 | 134.3 | 134.3 | 134.1 |
| Forest indust. | 59.0 | 61.4 | 67.6 | 68.9 | 70.6 | 71.5 | 71.5 | 71.4 | 71.3 | 71.0 |
| Farmer and other pvt. | 297.0 | 301.2 | 286.3 | 278.0 | 276.4 | 270.0 | 266.9 | 262.9 | 259.7 | 257.5 |
| Total | 508.8 | 515.1 | 504.1 | 491.1 | 483.2 | 475.8 | 472.7 | 468.6 | 465.2 | 462.6 |
| Region | | | | | | | | | | |
| North | 154.3 | 156.6 | 154.4 | 153.4 | 154.6 | 154.4 | 153.6 | 151.7 | 150.5 | 149.5 |
| South | 204.5 | 208.7 | 203.3 | 198.4 | 195.4 | 191.3 | 190.0 | 188.6 | 187.4 | 186.8 |
| Rocky Mt. | 66.6 | 66.9 | 64.5 | 60.2 | 61.1 | 59.9 | 59.7 | 59.5 | 59.4 | 59.2 |
| Pacific C. | 83.4 | 82.9 | 81.8 | 79.1 | 72.1 | 70.2 | 69.5 | 68.7 | 68.0 | 67.1 |
| Total | 508.8 | 515.1 | 504.1 | 491.1 | 483.2 | 475.8 | 472.7 | 468.6 | 465.4 | 462.6 |

*Source*: Waddell et al. (1989)

forest from abandoned farmland, though at a reduced rate.

From the point of view of afforestation it is thought that the South will be most affected by these changes. The timber industry is increasingly focusing its attention on this region and away from the once pre-eminent Pacific Northwest because the trees grow rapidly in the warm sub-tropical climate. Even so, there is likely to be a decline of total forest from 195 million acres (78.9 Mha) in 1987 to 187 million acres (75.7 Mha) by 2040, due largely to a fall in farm forest ownerships which are being bought up by an expanding lumber industry, which owned over 38 million acres (15.8 Mha) in 1987. These lands are likely to be afforested in pine plantations in order to allow genetically improved stock to be introduced and trees to be spaced to reduce management costs. It is probable that if current trends are continued the current *c.* 20 million acres (8.1 Mha) of plantations will rise to 45 million acres (18.2 Mha) and that the area of natural pine will fall accordingly (Alig and Wyant 1985; Fecso *et al.* 1982; USFS 1988; 1990).

In other regions afforestation is likely to be minimal. In the North and Northeast the natural succession of hardwoods on abandoned farmland will favour maple–beech at the expense of aspen–birch and oak–hickory associations. In the Pacific Coast region, particularly in the Pacific Northwest, the decline of forest area from 72.1 million acres (29. 2 Mha) in 1987 to 67.1 million acres (27.1 Mha) in 2040 will occur mainly in public ownerships as more land is reserved for other (recreational) uses, but it will be accompanied by little afforestation if company disinvestment continues as is thought likely (Robbins 1985; Le Master 1977).

All the indications are that the pressures of continued population growth and demand for forest products will put an increasing strain on the forest resources of the country. Unlike in the past, there will be less and less recourse to the vast stock of untapped forest, and increasingly afforestation, together with other techniques of raising production, will loom larger in the story of the Americans and their forests.

## References


Abernethy, Y. and Turner, R.E. (1987) U.S forested wetlands: 1940–1980, *Bioscience* 37: 721–7.

Alig, R.J. and Wyant, J.G. (1985) Projecting regional area changes in forestland cover in the U.S.A., *Ecological Modeling* 29: 27–34.

Aughey, S. and Wilbur, C.D. (1880) *Agriculture beyond the 100th meridian; or, a review of the United States Public Land Commission.* Journal Company, Lincoln, NB.

Birch, B.P. (1974) 'Initial perception of prairie: an English settlement in Illinois', in R.G. Ironside (ed.) *Frontier Settlement,* University of Alberta Studies in Geography, Monograph 1. Calgary, pp 178–98.

Clark, T.D. (1984) *The greening of the South: the recovery of land and forests.* University Press of Kentucky, Lexington.

Clawson, M. (1979) Forests in the long sweep of American history, *Science* 204: 1168–74.

Clawson, M. (1981) Competitive land use in American forestry and agriculture, *Journal of Forest History* 25: 222–7.

Dana, S.T. and Fairfax, S.K. (1980) *Forest and range policy: its development in the United States.* McGraw-Hill, New York.

Droze, W.H. (1977) *Trees, prairies and people: a history of tree planting in the Plains States.* Texas Woman's University Press, Denton.

Emmons, D.M. (1971a) American myth: Desert to Eden: theories of the increased rainfall and the Timber Culture Act of 1873, *Forest History* 15: 6–14.

Emmons, D.M. (1971b) *Garden in the grasslands: boomer literature of the Central Great Plains.* University of Nebraska Press, Lincoln.

Fecso, R.S., *et al.* (1982) *Management practices and reforestation decisions for harvested Southern pinelands.* Staff Report AGE 5821230, Washington, DC: US Department of Agriculture, Statistical Reporting Service.

Greeley, W.B. (1925) The relation of geography to timber supply, *Economic Geography* 1: 1–11.

Hart, J.F. (1968) Loss and abandonment of cleared farmland in the eastern United States, *Annals, Association of American Geographers* 58: 417–40.

Hart, J.F. (1980) Land use change in a Piedmont County, *Annals, Association of American Geographers* 70: 492–527.

Hayden, F.V. (1867) 'The geology of Nebraska. Report of the Commissioner of the General Land Office', in US Department of the Interior, *Annual Report for 1867*, 40th Congress, 3rd session, H.Ex. Doc. 1 (Serial no. 1326). GPO, Washington, DC.

Hendrickson, C.I. (1933) 'The agricultural land available for forestry', in US Congress, Senate, *A national plan for American forestry* [Copeland Report], GPO, Washington, DC.

Hibbard, B.H. (1965) *A history of the public land policies.* Reprint by University of Wisconsin Press, Madison.

Jordan, T. (1964) Between the forest and the prairie, *Agricultural History* 38: 205–16.

Kollmorgen, W.M. (1969) The woodman's assault on the domain of the cattleman, *Annals, Association of American Geographers* 59: 215–39.

Kollmorgen, W.M. and Kollmorgen, J. (1973) Landscape meteorology in the Plains Area, *Annals, Association of American Geographers* 63: 424–41.

Le Master, D.C. (1977) *Mergers amongst the largest forest products firms, 1950–1970*. Washington State University, College of Agricultural Research, Bulletin no. 854, Pullman, WA.

MacDonald, P.O., Frayer, W.G., and Clauser, J.K. (1978) *Documentation, chronology, and future projections of bottomland hardwood habitat loss in the Lower Mississippi Alluvial Plain*. Fish and Wildlife Service Division, Ecological Services, Vicksburg, MS.

McIntosh, C.B. (1975) The use and abuse of the Timber Culture Acts, *Annals, Association of American Geographers* 65: 347–62.

McManis, D.R. (1964) The initial evaluation and utilization of the Illinois prairies, 1815–1840. University of Chicago, Department of Geography Research Paper, no.94.

Marsh, G.P. (1848) 'Address before the Agricultural Society of Rutland'. Quoted in D. Lowenthal, On the author of man and nature, *American Forests* 83 (1977): 6, 8–11, 44–8.

Olsen, J.C. (1972a) Arbor Day: a pioneer expression of concern for the environment, *Nebraska History* 53: 1–13.

Olsen, J.C. (1972b) *J. Sterling Morton*. Nebraska State Historical Society Foundation, Lincoln.

Overfield, R.A. (1979) Trees for the Great Plains: Charles E. Bessey and forestry, *Journal of Forest History* 23: 18–31.

Raney, W.F. (1919) The Timber Culture Acts, *Mississippi Valley Historical Association Proceedings* 10: 219–29.

Reuss, L.A., Wooten, H.H. and Marschner, F.J. (1949) *Inventory of major land uses*. United States, USDA, Miscellaneous Publication no. 663, GPO, Washington, DC.

Risbrut, C.D. and Elleson, P.V. (1983) An economic evaluation of the 1979 Forestry Incentives Program. *Station Bulletin 550–1983*, St Paul, MN : University of Minnesota, Agricultural Experimental Station.

Robbins, W.G. (1985) The social context of forestry: the Pacific Northwest in the twentieth century, *Western Historical Quarterly*, 16: 412–23.

Royer, R. P. and Moulton, R.J. (1987) Reforestation incentives – tax incentives and cost sharing in the South, *Journal of Forestry* 85: 45–7.

Salmond, J.A. (1967) *The Civilian Conservation Corps, 1933–1942: a New Deal case study*. Duke University Press, Durham, NC.

Smith, H.N. (1947) Rain follows the plow: the notion of increased rainfall for the Great Plains, 1844–1880, *Huntingdon Library Quarterly* 10: 169–93.

Smith, H.N. (1950) *The virgin land: the American West as symbol and myth*. Harvard University Press, Cambridge, MA.

USBC (US Bureau of the Census) (1920) *Fourteenth census of Agriculture and 'Analytical Tables'*, pp. 33–47.

USBC (1977) *Historical statistics of the United States from Colonial Times to 1957*. Published in conjunction with the Social Science Research Council. GPO, Washington, DC.

US Congress (1872) *Congressional Globe*, 42nd Congress, 10th June 1872, pt 5, 4464.

US Congress, House (1890) *Bill to repeal the Timber Culture Laws and for other pur-*

*poses*. 51st Congress, 1st session, 1890, H. Res. 7254.

US Department of the Interior (1884) *Annual report for 1883*, pp 7–8.

USFS (US Forest Service) (1920) *Timber depletion, lumber prices, lumber exports and concentration of timber ownership* (Capper Report). U.S.Congress, Senate, 66th Congress, 2nd session, report on Resolution 311. GPO, Washington, DC, p 3.

USFS (1982) *An analysis of the timber situation of the United States: 1952–2030*. Forest Resources Report 23. US Department of Agriculture, Forest Service, Washington, DC.

USFS (1988) *The south's fourth forest: alternatives for the future*. Forest Resources Report 24. USDA, Forest Service, Washington, DC.

USFS (1990) *An analysis of the timber situation in the United States: 1989–2040*. General Technical Report RM–199. GPO, Washington, DC.

Vance, R.B. (1932) *The human geography of the South: a study in regional resources and human adequacy*. University of North Carolina Press, Chapel Hill.

Vaughan, L.M. (1928) Abandoned farm areas in New York. *Cornell Agricultural Experimental Station Bull. no.490*: Ithaca, NY.

Waddell, K.L., Oswald, D.D. and Powell, D.S. (1989) Forest statistics of the United States, 1987. Resource Bulletin. PNW–RB–168. U.S.Department of Agriculture, Forest Service, Pacific Northwest Research Station, Portland, OR.

Wessel, T.R. (1967) Prologue to the Shelterbelt, 1870–1934, *Journal of the West* 6: 119–34.

Williams, M. (1989) *Americans and their forests*. New York, Cambridge University Press.

Williston, H.L. (1979) *A statistical history of tree planting in the South, 1925–79*. USDA, Forest Service, Atlanta.

Wilson, H.F. (1936) *The hill country of northern New England: its social and economic history, 1790–1830*. Columbia University Press, New York.

Wilson, J. (1868) 'Observations ... on forest culture'. Report of the Commissioner of the General Land Office, 1868, in US Department of the Interior, *Annual report for 1868*, 40th Congress, 3rd session, H.Ex Doc. 1 (Serial no.1366). GPO, Washington, DC, pp 173–98.

# 14 Review

*Alexander Mather*

The national experiences of afforestation discussed in this volume are extremely varied. In Europe and the United States, large areas of land that were treeless one hundred years ago are now forested, and similar processes are now underway in a number of other parts of the world.

In some cases, for example in the countries of north-west Europe, the native forest had almost completely disappeared by the eighteenth or nineteenth centuries, and most of the present forest area is the product of afforestation. Elsewhere, for example in parts of the New World, afforestation and reforestation began while substantial areas of native forest survived. In such cases afforestation has accompanied deforestation. It remains to be seen whether the overlap in time of these two processes will be a temporary phenomenon, or whether it will persist indefinitely.

Each country here reviewed has its own unique history of afforestation policy, planning and progress. Nevertheless, many countries share certain similarities, for example in terms of evolutionary trends in afforestation policies and of social and environmental effects. Different criteria can be employed in assessing 'progress', and different evaluations can be made of the relative costs and benefits of afforestation policies. It is clear that no single country can serve as a perfect afforestation model for others to emulate. At the same time, however, the evolutionary trends of countries with long traditions of afforestation, and the criticisms focused on the afforestation policies of such lands, may be of interest in countries that have more recently adopted afforestation policies.

## Does planting trees save the forests?

This book began with a statement to the effect that planting trees would in itself not save the forests (of the former Soviet Union), but it was a good beginning. If the statement is valid for the former Soviet Union, it is perhaps also applicable at the global scale. Certainly it is difficult to see the global forests being saved if trees are *not* planted. An afforestation policy is, of course, no substitute for an effective and comprehensive forest policy. It does not in itself offer a solution or a 'technical fix' to the problem of shrinking forest resources. Indeed it has been suggested that 'reforestation ... addresses the symptom but not most of the underlying causes of non-sustainable exploitation of forest resources' (Kunstadter 1990, p 187). In many parts of the world, however, and in some of the countries discussed in this volume, the original forest resource has been severely depleted, almost to the point of exhaustion. In such settings it is obvious that afforestation is likely to feature prominently in any forest policy adopted in such a land. It is too late there to save the (original) forest. The best that can be done is to create new forests, and to offset the shrinkage of the forests in some parts of the world with expansion in others.

In some instances, the creation of forest plantations can destroy native forests, rather than helping to save them. Indeed, the establishment of commercial tree plantations in Thailand has been described as 'deforestation by any other name' (Lohmann 1990). The replacement of natural or semi-natural woodlands by plantations has not been confined to the tropics: the 'coniferisation' of such broadleaved woodlands continued to some extent in Britain until the 1980s, and in (West) Germany the percentage of conifers in the growing stock increased from 30 at the end of last century to 70 by 1960 (Zobel *et al.* 1987). The process of replacement of native woodlands by plantations has been prominent in countries such as Chile. As the process of afforestation extends into the tropics, it is clear that it carries with it some threat to remaining areas of natural forest.

Yet this threat should not be exaggerated. In countries such as Britain, remaining areas of natural or semi-natural woodland are now protected more effectively than ten years ago, even if the protection is still imperfect. In the tropics, Evans (1982) estimated that around 30 per cent of plantations had been established in areas of natural forest and woodland, but ten years later he concluded (Evans 1992) that the proportion was 'now even less'. He reports that many tropical afforestation projects are now specifically targeted at degraded or waste land. Nevertheless, some threat remains, and the Tropical Forest Action Plan produced by FAO has attracted criticism from commentators such as Ross and Donovan (1986) and Shiva (1987) for its possible effects in assisting the conversion of natural forests to plantations, rather than tackling the deeper roots of deforestation. Extensive areas of grassland and waste land occur in tropical latitudes, often in an 'under-used' state (Evans 1992). The fact that such areas may be capable of afforestation highlights the apparent wastefulness of converting remaining areas of tropical forest to plantations.

Even if plantations are not directly replacing natural forest, afforestation can have negative effects. Afforestation policies may become the focus of forestry policy, and distract attention from the need for active conservation – as in the case of Algeria (Chapter 7).

If it is true that under certain circumstances tree planting can be the enemy of the forest, it is also the case that the forest will be degraded or will disappear completely unless trees are planted. Perhaps recent popular concern about the fate of the tropical forest has diverted attention from the condition of the boreal coniferous forests in Canada and the (former) Soviet Union. Huge areas of land – amounting in the Soviet Union to 138 million ha – require replanting. For many years the rate of replanting was much less than that of logging (Gillis and Roach 1986; Braden 1991). Planting and reseeding rates have accelerated in recent years in these areas, but large deficits remain. Perhaps in the case of the former Soviet Union, it may 'be too late for true forest preservation in harvested regions', and the result of planting may be 'new tree farms, not renewed forest ecosystems' (Braden 1991, pp 134 and 133). The alternative without planting, however, may simply be degraded and unproductive scrub land.

It is frequently argued that afforestation can also help to save the forests in an indirect way. Because of their high productivities, industrial and fuelwood plantations can produce large quantities of wood from small areas of land. Pressures on remaining areas of native forests and woodlands may therefore be lessened, at a variety of scales ranging from the local to the international. It is perhaps significant in this respect that some conservation bodies, such as the Australian Conservation Foundation, support plantation programmes for this reason (Chapter 11). Wood production, however, is usually only one of the pressures on the native forest. If clearance for agricultural purposes or for other land uses is the main threat, then it is obvious that afforestation will not in itself preserve the forest area. It is noticeable that forest clearance is continuing in Australia, for example, despite an afforestation programme geared to wood production and a plethora of schemes to encourage planting for environmental reasons.

Furthermore, domestic afforestation within the bounds of a country does not necessarily result in decreasing imports of wood from natural forests elsewhere. For reasons such as type, quality and cost of wood, imports may continue even after extensive afforestation or reforestation, as the case of Japan in particular illustrates.

## The introduction of afforestation policies

Afforestation policies have been introduced in a wide range of historical and political circumstances. By definition, state policies involve some form of government intervention. Individual land owners can and do carry out afforestation without or outside national policies, but such efforts are usually short-lived, except, perhaps in the case of the replanting of industrial plantations by corporate owners. In many of the countries reviewed in this volume, the state was instrumental in initiating afforestation, either directly on land acquired by it for that purpose, or indirectly through incentives offered to the private sector. The afforestation of bare land is rarely an attractive financial proposition to the private individual, and some form of financial incentive, such as planting grants or tax concessions, is usually required if significant amounts of new planting are to be achieved. Equally, corporations are rarely likely to replant harvested 'old growth' land unless sanctions or incentives are provided, or unless no old-growth forests remain to be exploited.

### *Active and passive factors*

The timing of the introduction of afforestation policies is not easily explained, but is probably determined by a combination of active and passive factors. The latter may include a keen perception of the disadvantages of depleted forest resources and possession of the technical, financial and organisational ability to carry out afforestation. That these conditions may be necessary but not sufficient to trigger afforestation is illustrated in microcosm by the case of St Helena. There the environmental effects of deforestation were recognised from the eighteenth century, and both temperate and tropical tree species were first introduced in that same century. Yet large-scale afforestation did not begin until the second half of the present century (Barlow 1989). In the case of countries such as Britain the interval between, on the one hand, the recognition of the problems and the availability of suitable species and technology, and on the other hand the advent of effective afforestation policy, was even longer.

Active factors frequently involve the stimulus of actual, perceived or feared wood shortages. This stimulus has operated in a variety of time periods and geographical settings. Britain is a classic example, where the short-term crisis of wood scarcity during World War I was superimposed on the long-term shortages of indigenous and Empire supplies. The short-term crisis was sufficient to trigger an afforestation effort that has persisted for three-quarters of a century. A stimulus of a kind different in form but similar in effect was the change in national boundaries and loss of forest land in Hungary after World War I. The spirit of a newly independent nation-state, seeking to make maximum use of the land resources now under its control, may be a similar if less dramatic stimulus.

At a different level, the role of key personalities in contributing to the establishment or expansion of afforestation programmes at such times seems significant if not crucial. Examples in these pages include Connolly and McBride in Ireland, Kaán in Hungary and Monjauze in Algeria. The conjunction of the personality, the period and the opportunity may be hard to predict, but there seems little doubt that some individuals emerged at particular periods of opportunity and steered their countries towards forest expansion. Political upheavals in particular appear to have been congenial climates for such individuals to emerge.

Stimuli or trigger factors may be sudden and dramatic, but without the appropriate political and economic climate might be quite ineffective. In the case of Britain, for example, it is, with hindsight, clear that the climate had becoming progressively more favourable for afforestation during the twenty or thirty years before the afforestation policy was formally adopted. This 'pre-afforestation' era helped to prepare political and public opinion for the radical measures that were soon to be adopted. Perhaps the outcome of World War I, in terms of afforestation policy, might have been different but for the developments and evident changes in attitudes around the turn of the century.

### *The quest for self-sufficiency*

The quest for increased or complete self-sufficiency in forest products has perhaps been the most common reason for the adoption of afforestation policies and policies of forest expansion. It has been pursued around the world, from Australia (Chapter 11) to Venezuela (Fahnstock *et al.* 1987). Increased self-sufficiency has been sought for both strategic and for economic reasons, and in the face of acute short-term problems of wood supplies and of fears of long-term shortages.

Fuelwood shortages are, of course, acute in some parts of the world, and some industrial wood shortages have been real enough in other areas (not least in China). Many predictions of wood shortages, however, have not materialised. In some cases the conditions giving rise to the gloom were changing even when the predictions of national shortages were being made: the United States at the beginning of the twentieth century is the classic example. In other cases, the predictions of shortages were based on demand forecasts that proved to be unrealistic, either because per capita demand turned out to be lower than predicted or because population growth rates declined. In countries as diverse as Britain, New Zealand, the United States and Australia, many of the predictions of shortages that encouraged thought about afforestation have not occurred. In Australia, for example, Carron (1980) observes that the 'scarcity' argument has been used for seventy years without the scarcity coming to pass.

In general terms, the failure of predictions of scarcity may stem from a variety of reasons: the failure to predict accurately trends in demand; the failure to foresee changing circumstances; and possible exaggeration of likely scarcity in the first place, in order to expand or maintain programmes of forest expansion.

It is possible that predictions of wood scarcity at the national level may play a lesser part in the future than in the past. Wood and timber are clearly less important strategic materials than they were in the days of sailing ships, and probably also less important than at the time of World War I. And with the increasing internationalisation of the forest-products industry, considerations of comparative locational advantages in the costs of wood production may become increasingly important. If so, the perceived need for individual countries to seek increased levels of self-sufficiency may lessen.

### *Environmental protection*

While increased self-sufficiency in wood supply, for economic or strategic reasons, may have been a major factor in the initial formulation of many national afforestation policies, it is by no means the only one. Environmental protection may be another major objective, and is one that has been sought from the earliest days of afforestation, as, for example, in the case of planting on some European coastal sand dunes from medieval times onwards (Chapter 1). There have been cases of serious misperceptions of the environmental relations of afforestation, and of misguided policies. Afforestation on the grasslands of the United States and the 'Green Dam' scheme in Algeria are examples (Chapters 13 and 7, respectively). On the other hand, environmental improvements have frequently resulted, as in the case of Hungary (Chapter 5), and there is no doubt that a new drive for afforestation for environmental reasons is now underway in some parts of the world, as the case of Australia demonstrates (Chapter 11). If scarcity of wood can trigger a quest for greater self-sufficiency and hence afforestation, then perhaps environmental problems may have a similar effect.

## The content and evolution of afforestation policies

Policies of forest expansion often begin with simple, clearly defined objectives, perhaps set in the light of immediate problems such as wood shortage. With the passing years, however, new objectives are often added, so that the goals become more blurred and uncertain. Policy evaluation may become more difficult, and the priorities of the various objectives are not always identified, nor are the criteria by which they may be assessed always defined.

Typically, policies begin with objectives of increased wood production, perhaps combined with social benefits such as the provision of rural employment. Environmental benefits and the provision of recreational opportunities are often added subsequently. For example, in Britain the initial primary objective of the policy of forest expansion was increased wood supply, combined with a secondary objective of providing rural employment. Recreation was added, initially in the 1930s and more especially in the 1960s, and environmental considerations have attracted increasing attention in subsequent decades.

The extent to which objectives other than wood production feature in policies of afforestation varies from country to country. The pages of this volume suggest that they are most prominent in the countries with the oldest-established policies, and that an evolutionary trend towards complexity or multiplicity of objectives can be discerned. On the other hand, examples such as Australia suggest that environmental objectives may emerge even in some countries that have begun to afforest only recently. Processes of evolution or maturing of policies that took more than half a century in European countries, may, if the case of Australia is typical, operate more rapidly in lands recently adopting afforestation.

Increasing complexity of objectives may be reflected either in the increasing complexity of grant schemes, or in a multiplicity of different schemes. In either case, the small land owner may become more confused and less responsive to the incentives on offer. Periodic attempts may be made to streamline or rationalise grant schemes, which at least in the short term can add to the confusion and lessen the response. Furthermore, if the objectives of policies vary through time, their *intensity* may also fluctuate. Target areas for planting and real values of incentives can be rapidly adjusted, depending on the perceived significance of predicted wood shortages, unemployment levels or the need for environmental protection. The lack of consistency through time of objectives and targets is a major theme emerging in the preceding pages. Although trees are usually a long-term crop, afforestation policies are often subject to short-term change and volatility, not least in relation to the changing economic and political climate of a country. Sudden changes in planting targets and incentives are quite common: few governments are keen to give long- or even medium-term commitments. Richardson (1990) regards five-year planning periods as unsatisfactory from the viewpoint of afforestation, as they are too long to sustain the enthusiasm of planting campaigns and too short to provide an orderly expansion of forest resources. His conclusion was based on China, and on the type of five-year planning typical of the communist world. But such planning horizons are common also in the West: in Britain during the 1950s and 1960s, for example, five-yearly forest policy reviews were carried out. Such timescales fit well with political planning horizons, but less well with an activity such as afforestation. Time lags usually occur between the introduction of new measures such as planting-grant schemes and their widespread uptake by afforesters. Frequent changes in the objectives of policies, and in their instruments, can thus result in a rather unsteady or spasmodic pattern of afforestation, with obvious implications for future supplies of wood.

Objectives other than wood production vary between sectors in individual countries as well as between different countries. State afforestation can obviously incorporate objectives such as environmental protection and public recreation more easily than that on private land. On private land, additional incentives are usually required if progress is to be achieved in afforestation for recreational or environmental purposes, over and above those needed to stimulate afforestation for wood production. As state-sector afforestation lessens in importance around the world, relative to that in the private sector, the significance of such incentives will increase. For progress towards the desired objectives to be achieved increasing care will be required in designing and targeting the incentives. Numerous examples – including in these pages settings as diverse as Britain and Chile – can be cited of incentives that, intentionally or unintentionally, favour one type of land owner or potential afforester rather than another. While rapid rates of afforestation may have been achieved, social and environmental costs may have been incurred and potential benefits unrealised.

The increasing emphasis on private-sector afforestation also raises the question of planting targets. When afforestation is carried out mainly by the state, then targets can in theory be achieved relatively easily, though in some cases the lack of harmonisation between different sectors of land policy may pose problems. The achieving of national targets by the private sector is more difficult. The design of packages of incentives is rarely sufficiently sophisticated to achieve a precise result in terms of annual planting targets.

Planting targets are in any case of debatable value and origin. In some cases, it seems that their attainment can become an objective or end in itself, with scant consideration for the basis or justification of the target figure. The translating of general objectives into precise targets is often a rather arbitrary process: in some cases, perhaps, the targets amount to no more than generalised signals for expansion rather than goals for precise achievement. If wood production is a primary objective, areal targets need to be set in the context of land productivity, and if recreational or environmental benefits are sought then the locational characteristics of the new or expanded forests are of major significance. In other words, crude national targets for afforestation may be of limited practical importance. In the years ahead, perhaps targets more disaggregated in terms of objectives, types and locations of planting will replace the crude national targets that have been set – though not always achieved – in numerous countries around the world.

## The limits to afforestation policies

As the previous chapters have made clear, progress in afforestation is not a matter of afforestation policies alone. Agricultural policies may be at least as important. Forestry is usually subordinate to agriculture in national priorities and in the influence of lobbies and pressure groups, and what happens in agriculture may to a large degree dictate the scale, rate and pattern of afforestation, whatever forestry policies indicate. The desired release or protection of agricultural land, for example, may have a major bearing on achievements in afforestation. For example, forestry was viewed as an alternative use for surplus or degraded agricultural land in the United States during the 1930s, and both in the United States and in many other parts of the developed world since the second half of the 1980s. Agricultural circumstances may influence not only the amount of land available for afforestation, but also its type and pattern and the type of incentives offered to afforesters.

Equally, agricultural policies may thwart afforestation policies. In post-World War II Europe, for example, the need to protect food-producing land was seen as a priority, and the achievement of afforestation targets was sometimes frustrated and afforestation was relegated to residual areas of poor-quality land. It is by no means always the case that agricultural and forestry policies are integrated or harmonised. For successful progress to be achieved in afforestation, the integration of afforestation policies with wider forestry, agricultural and development policies appears essential.

## Afforestation: social effects

Afforestation has been widely perceived as offering or providing social benefits, especially in terms of higher intensity of rural employment than the previous land uses supported. Some areas have indeed benefited in this way, and more especially from additional employment in transport and wood processing. Even if employment levels in the new forests and on alternative land uses such as sheep farming are comparable, the former may employ several times more workers per unit area than the latter when downstream employment is considered. In Otago, New Zealand, for example, the ratio is reported to be more than four to one in favour of forestry when downstream employment is included, whereas there is approximate parity when primary employment alone is considered (Aldwell and Whyte, 1984). At the other end of New Zealand, in Northland, afforestation programmes were found to arrest and reverse trends of rural depopulation, and to help to create more diverse local communities with higher percentages of young married couples and hence better-supported local services (Farnsworth 1983).

On the other hand, a tension often exists between social and economic objectives, and the social experience has sometimes been disappointing. In Australia, for example, it was forecast that employment in the forest industry would rise substantially, whereas it has actually fallen (Dargavel 1982). In Portugal, some eucalypt plantations have been found to support lower employment densities than the olive groves and vineyards that they replaced (e.g. Kardell *et al.* 1986).

Experience has sometimes also been disappointing in terms of the type of employment provided, as well as of the amount. Large-scale afforestation, especially where carried out by corporations or absentee owners, may involve mobile squads of workers, rather than local residents. While the peaked nature of labour demand may favour such a pattern of employment, the local social and economic benefits may be considerably less than if local residents were employed. This in turn raises the question of the nature of incentives for afforestation, and the type of afforestation that they encourage. Relatively few national examples can be quoted where most of the afforestation has been carried out by small land owners or local community groups. In countries such as Britain the latter have benefited relatively little from afforestation policies, and in lands such as Spain and Chile they have suffered costs of displacement or loss of land. The result is at best a lack of involvement of local people in afforestation, at worst complete alienation.

One of the major themes emerging from the preceding chapters is the 'top-down' nature of many afforestation policies, and the way in which they have been imposed on, rather than being devised with, local communities. A related issue is the integration of national policy and local and regional planning. The regional allocation of national planting targets, for example, has not always been harmonised with the priorities of such planning, and uneasy relationships between planning authorities, potential afforesters and the state forestry service are by no means unknown. In recent years, however, there are clear signs of change in this respect in countries such as Britain, Denmark and Spain. The role of planning, and indirectly of the public, in influencing the location and type of afforestation is growing. In addition, more thought has been given to providing policy instruments that will encourage farmers and small land owners to participate in afforestation. The main deterrent has been the long interval between initial investment and eventual harvesting of the tree crop. Over the last few years new schemes offering annual payments have been introduced. Planting under these schemes may be small both in terms of individual plots and of aggregate areas, and their contribution to national afforestation may be limited. Nevertheless, their significance in terms of environmental and landscape effects may be greater.

## Afforestation: environmental effects

Historically, tree planting has been viewed as an environmentally beneficial and benign activity. Indeed, environmental protection has been an objective of afforestation for several centuries, and environmental benefits have been achieved in countries as diverse as the United States and Hungary. Afforestation has up to now usually been associated with environmental benefits on the local or regional scales, for example through control of soil erosion or flooding. Recently, however, possible benefits at the global scale have been identified in the form of reduction in atmospheric carbon dioxide (e.g. Sedjo 1989a; 1989b). Parts of the world where deforestation is proceeding rapidly (such as Latin America) are major sources of carbon dioxide (e.g. Houghton *et al.* 1991). Other areas that previously suffered deforestation and have now been partly reforested have changed from sources to sinks. One major example is the American South, where the rapid growth of forest plantations means effective uptake of carbon dioxide (Delcourt and Harris 1980). Afforestation on a massive scale would be required if a significant effect were to be achieved globally, but some individual projects suggest that the carbon dioxide issue and the associated possibility of global warming may boost afforestation efforts. One American power company, for example, is reported to have agreed to plant 52 million trees on a 500 $km^2$ area in Guatemala, in the hope that they will absorb at least as much carbon dioxide as its new power station in Connecticut (Tyler 1990).

While potential environmental benefits have long been recognised, more recently afforestation has sometimes been opposed or resented on the grounds of its adverse environmental impact, especially in some European countries. This negative perception has stemmed from two main sources – the type of preparation and management practices employed, and the type of forests created.

The effects of ground preparation for afforestation on local hydrology and patterns of erosion and sedimentation are now well known, having been almost invariably experienced wherever large-scale afforestation has been carried out. Simple means are available for mitigating such effects, for example through the use of buffer zones along streams. Their use can be encouraged through the issuing of guidelines and the establishing of codes of practice. If the political will is present, their use can be encouraged more strongly through conditions imposed on planting grants or by other stringent means. Similarly, the careful and sympathetic design of new forests can help to minimise adverse impact on landscape and wildlife. There is no doubt that many mistakes were made in these respects in the early days in the countries with longer histories of afforestation. Perhaps rapid or ambitious afforestation programmes, where there is more emphasis on the amount of afforestation carried out than on its quality, are especially prone to such problems. Unfortunately, popular attitudes towards afforestation are adversely coloured as a result.

Many national afforestation programmes around the world have focused on the use of a very small number of species, such as Sitka spruce in Britain and Ireland, eucalypts in Spain, Portugal and many low-latitude countries, and Radiata pine around the Pacific Rim. Such species have usually been selected for a combination of reasons, such as attractive growth rates and suitability for a wide range of site conditions. The use of near-monocultures of exotic species in new and expanded forests is widely criticised on a variety of grounds. One is the effect on wildlife, which is usually less diverse in exotic plantations than in native forests. In Australia, for example, species richness is lower and the proportion of introduced species higher in pine plantations than in native eucalypt forests (Friend 1982). Faunal diversity has been found to increase with age of trees in New Zealand pine plantations, but rarely if ever becomes as high as in unmodified native forests on similar sites. On the other hand, such plantations have been found to provide useful habitats for some native animal species (Bull 1981). Wildlife has usually been found to increase when exotic plantations are established on badly degraded grassland or scrub (Zobel *et al.* 1987). The risk of fire and outbreaks of pests and diseases may also be greater in exotic plantations than in natural forests or in plantations of native species. Such outbreaks have occurred in settings as diverse as Britain, Chile and New Zealand.

In response to criticisms of excessive reliance on a very small number of species, some diversification has become apparent in recent years in countries such as Ireland, Spain, Chile and Britain. Where afforestation is carried out under state influence, for example with the assistance of planting grants, conditions on species mixes and proportions can be imposed.

Another criticism of conventional plantations is that while productivity may be high during the first rotation, it may decline thereafter. This decline may stem either from management practices at the time of felling and extraction, or from longer-term drain of nutrients from the soil. Evidence from around the world is conflicting: in some countries declines have been reported, in others no decrease has been observed and in others again declines have been noted in some plantations but not in others. In short, the question of sustainability is uncertain.

The other major criticism of reliance on a very small number of species, made for example in the chapter on Spain, is that plantations of exotic species offer only industrial wood and then only for processing or for non-specialised purposes: the wide variety of products that can be obtained from natural forests or for some other kinds of plantations are not available.

## Industrial and post-industrial forests

Most of the plantations established during the twentieth century have been geared to the production of wood for industrial purposes. As the century draws to an end, there are signs that such forests are becoming increasingly unwelcome, especially in western Europe. As Chapter 2 indicates, commercial afforestation using coniferous species has in effect been banned from the uplands of England because of its unacceptability on environmental grounds.

Whereas wood production enjoys complete primacy in the industrial forest, there is now growing popular demand for new forests to serve multiple purposes. In some settings, such as around large cities, objectives other than wood production may now achieve primacy. Even if wood production remains a major objective, it may have to be combined with other objectives, and may be constrained by the types of management practices and species composition deemed acceptable. This shift from the 'industrial' to the 'post-industrial' forest has in Europe coincided with changes in the type of land available for afforestation. Up until the mid-1980s, only the poorest qualities of land were usually released for planting. Now, with over-production of some agricultural commodities, better qualities of land are available, and may support more diverse forms of afforestation. In particular, far more broadleaved trees are likely to be planted.

These trends are potentially of profound significance for afforestation at a variety of spatial scales. The 'push' factor of opposition to 'industrial' afforestation in European countries and the 'pull' factors of the rapid growth rates and short rotations possible in sub-tropical or tropical lands may mean that 'industrial' afforestation is increasingly concentrated in low latitudes. The trends of internationalisation highlighted in the chapter on New Zealand may simply facilitate such a tendency. Economic imperatives may themselves mean that the tropics and sub-tropics will enjoy comparative advantage. It is perhaps significant that corporations in Chile are at present able to carry out commercial afforestation without grant aid (Chapter 9). The 'push' of environmental and planning constraints in the 'North' may simply serve to emphasise an already emerging trend. The environmental and social costs of such afforestation in countries such as Chile are already recognised: it is possible that these costs may intensify if constraints and safeguards are not incorporated. If they are in due course incorporated, perhaps 'industrial' afforestation will simply move to other lands offering lower costs and fewer controls.

Meanwhile, in European-type countries 'post-industrial' afforestation may proceed with government support, with greater use of broadleaved species than in the past and on better land, and with the pursuit of multiple objectives.

In an ideal world perhaps afforestation everywhere would be along these lines. Perhaps one of the general lessons to be learned from the experiences of European countries is that the typical 'industrial' type of afforestation may eventually become untenable in terms of popular acceptance and support. Unfortunately 'post-industrial' afforestation, offering a wider range of social and environmental benefits, may be more expensive initially, and take longer to yield returns. Private investment is likely, therefore, only if encouraged by more generous incentives from government than would be required to encourage traditional 'industrial' afforestation. It remains to be seen whether the newer afforesting countries will follow the 'European' path, and probably eventually suffer the same type of reaction as recently experienced in some European countries, or aim more rapidly for 'post-industrial' afforestation.

## Afforestation: research agenda

It is probably true to say that a far greater research effort has been directed at the technical aspects of silviculture than at the political, social, and environmental aspects of afforestation. Numerous questions and uncertainties exist about fundamental issues. How and why do afforestation policies originate? How do they relate to broader forestry policies, and to overall land policies? What are the determinants of their performance, and by what criteria are their progress and success to be evaluated?

One of the most basic questions concerns the determinants of the forest transition and how the adoption of afforestation policies relates to that transition. The dearth of reliable and comprehensive statistical data is a major problem: perhaps the only way forward in the search for greater understanding is the careful and painstaking assembling of time-series data for individual countries, so that eventually fuller international comparison and analysis becomes possible.

At a different level, the growing internationalisation of forest-based industry poses profound questions about patterns and styles of afforestation in the medium and long term. Most of the national afforestation programmes reviewed in these pages originated in domestic settings. In an age of multinational corporations and mobility of capital, the climate for afforestation may be quite different. As in other spheres of activity, individual governments and their ministries and state forest services may be no match for multinational corporations as they seek to identify and exploit the optimal environments for wood production. A greater understanding of these trends of internationalisation and globalisation is therefore urgently required.

Related to these macro-scale trends is the question of the incentives offered by governments for private afforestation. Recent years have witnessed a retreat of the state from direct involvement in afforestation, and a correspondingly increased importance for the private sector. For most afforestation, and especially for 'environmental' planting, financial incentives are essential. The nature of these incentives is of the utmost significance. By accident or design, they may favour certain types of potential afforesters, and they may have also have strong and perhaps unanticipated spatial effects. Furthermore, responses to them, in terms of uptake and hence of planting achievement, are often unpredictable. Again, a much fuller understanding of the characteristics of different types and packages of incentives is required.

The experiences of afforestation as reviewed in these pages are extremely diverse. The settings for the policies of afforestation, the ways in which the policies were implemented, and the degree of progress achieved are very varied. Sweeping generalisations are probably as unhelpful as simple conclusions are unrealistic. What is clear, however, is that many very different countries have recognised that their forests are too small or degraded for the needs and aspirations of their peoples to be satisfied or achieved. They have thus taken steps to increase the degree of tree cover. Rates of deforestation may now be at record levels. Rates of afforestation have accelerated even more spectacularly in the present century.

## References


Aldwell, P.H.B. and Whyte, J. (1984) Impacts of forest-sector growth in Bruce County, Otago: a case study, *New Zealand Journal of Forestry* 29: 269–95.

Barlow, A.R. (1989) Forestry development on the island of St Helena, *Commonwealth Forestry Review* 68: 57–68.

Braden, K. (1991) 'Managing Soviet forest resources', in P.R. Pryde, *Environmental management in the Soviet Union.* Cambridge University Press, Cambridge and New York, pp 112–134.

Bull, P.C. (1981) The consequences for wildlife of expanding New Zealand's forest industry, *New Zealand Journal of Forestry* 26: 210–31.

Carron, L.T. (1980) Self sufficiency in forest policy in Australia, *Australian Forestry* 43: 203–9.

Dargavel, J. (1982) Employment and production: the declining forestry sector examined, *Australian Forestry* 45: 255–61.

Delcourt, H.R. and Harris, W.F. (1980) Carbon budget of the Southeastern U.S. biota: an analysis of historical change from source to sink, *Science* 210: 321–3.

Evans, J. (1982) *Plantation forestry in the tropics.* Clarendon, Oxford.

Evans, J. (1992) *Plantation forestry in the tropics* (second edition). Clarendon, Oxford.

Fahnstock, G.R., Tarbes, J. and Yegres, L. (1987) The pines of Venezuela: can plantations supply domestic demands? *Journal of Forestry* 85(11): 42–4.

Farnsworth, M.C. (1983) The social impact of forest development in Northland, *New Zealand Journal of Forestry* 28: 246–54.

Friend, G.R. (1982) Mammal populations in exotic pine plantations and indigenous eucalpyt forests in Gippsland, Victoria, *Australian Forestry* 45: 3–18.

Gillis, R.P. and Roach, T.R. (1986) *Lost initiatives: Canadian forest industries, forest policy and forest conservation.* Greenwood Press, New York.

Houghton, R.A., Lefkowitz, D.S. and Skole, D.C. (1991) Changes in the landscape of Latin America between 1850 and 1985, *Forest Ecology and Management* 38: 143–200.

Kardell, L., Steen, E. and Fabiao, A. (1986) Eucalyptus in Portugal– a threat or a promise?, *Ambio* 15: 6–13.

Kunstadter, P. (1990) Impacts of economic development and population change on Thailand's forests, *Resource Management and Optimization* 7: 171–90.

Lohmann, L. (1990) Commercial tree plantations in Thailand: deforestation by any other name, *The Ecologist* 20(1): 9–17.

Richardson, S.D. (1990) *Forests and forestry in China.* Island Press, Washington.

Ross, M.S. and Donovan, D.G. (1986) The World Tropical Forestry Action Plan: can it save the tropical forests? *Journal of World Forest Resource Management* 2: 119–36.

Sedjo, R.A. (1989a) Forests to offset the greenhouse effect, *Journal of Forestry* 87 (7): 12–15.

Sedjo, R.A. (1989b) Forests: a tool to moderate global warming, *Environment* 31:14–20.

Shiva, V. (1987) Forestry myths and the World Bank, *The Ecologist* 17: 142–49.

Tyler, C. (1989) Towards a warmer world, *Geographical Magazine* 61(3): 40–3.

Zobel, B.J., Van Wyk, G. and Stohl, P. (1987) *Growing exotic forests.* John Wiley, New York.

# Index